Fuels and Petrochemicals Division

Held at the 2024 AIChE Annual Meeting

San Diego, California, USA
27-31 October 2024

ISBN: 979-8-3313-1658-7

Printed from e-media with permission by:

Curran Associates, Inc.
57 Morehouse Lane
Red Hook, NY 12571

Some format issues inherent in the e-media version may also appear in this print version.

Printed with permission by Curran Associates, Inc. (2025)

For permission requests, please contact AIChE
at the address below.

AIChE
120 Wall Street, FL 23
New York, NY 10005-4020

Phone: (800) 242-4363
Fax: (203) 775-5177

www.aiche.org

Additional copies of this publication are available from:

Curran Associates, Inc.
57 Morehouse Lane
Red Hook, NY 12571 USA
Phone: 845-758-0400
Fax: 845-758-2633
Email: curran@proceedings.com
Web: www.proceedings.com

Fuels and Petrochemicals Division

Held at the 2024 AIChE Annual Meeting

San Diego, California, USA
27-31 October 2024

AIChE
The Global Home of Chemical Engineers

TABLE OF CONTENTS

ADVANCES IN BIOFUELS PRODUCTION AND ALTERNATIVE FUELS I

ADVANCES IN PETROLEUM PRODUCTION AND REFINING I

DEVELOPMENTS IN ALTERNATIVE FUELS AND ENABLING TECHNOLOGIES

WASTE FEEDSTOCKS TO FUELS AND PETROCHEMICALS II

DEVELOPMENTS IN SHALE GAS AND NATURAL GAS

DEVELOPMENTS IN ALTERNATIVE FUELS AND ENABLING TECHNOLOGIES II

INTEGRATED BIOREFINERIES: TECHNOLOGIES AND TEA/LCA

POSTER SESSION: FUELS AND PETROCHEMICALS DIVISION

PROPERTIES AND PHASE EQUILIBRIA FOR FUELS AND PETROCHEMICALS

RECENT DEVELOPMENTS IN FUEL PROCESSING FOR HYDROGEN PRODUCTION I

VALUE-ADDED CHEMICALS FROM NATURAL GAS

WASTE FEEDSTOCKS TO FUELS AND PETROCHEMICALS I

ADVANCES IN BIOFUELS PRODUCTION AND ALTERNATIVE FUELS II

ADVANCES IN PETROLEUM PRODUCTION AND REFINING II

RECENT DEVELOPMENTS IN FUEL PROCESSING FOR HYDROGEN PRODUCTION II

RECENT DEVELOPMENTS IN FUEL PROCESSING FOR HYDROGEN PRODUCTION III

HARMONIZED COST AND AVAILABILITY ESTIMATES FOR EMERGING FUELS AND ENERGY CARRIERS

Doris Oke, Argonne National Laboratory, Lemont, IL, USA

Troy R. Hawkins, Argonne National Laboratory, Lemont, IL, USA

John Field, Oak Ridge National Laboratory, Oak Ridge, TN, USA

Ling Tao, National Renewable Energy Laboratory, Golden, CO, USA

Paige Jadun, National Renewable Energy Laboratory, Golden, CO, USA

Saurajyoti Kar, Argonne National Laboratory, Lemont, IL, USA

Krishna Reddi, Argonne National Laboratory, Lemont, IL, USA

Udayan Singh, Argonne National Laboratory, Lemont, IL, USA

Introduction

The long-term decarbonization strategy of the United States outlines the goal to reach economywide net-zero greenhouse gas (GHG) emissions by 2050 (U.S. Department of State and U.S. Executive Office of the President, 2021). Alternative energy carriers are near-indispensable to meeting this goal, contributing about 40% of emissions reductions. Achieving this target will require multifaceted changes to fuel production and conversion pathways across end-uses. As such, the cost and availability of low-carbon energy carriers derived from biomass will be key determinants of cost-optimal decarbonization pathways. Bioenergy pathways are characterized by differences in quality of feedstock, readiness of conversion pathways and competition with other alternatives. Additionally, availability and cost of bioenergy such as biofuels, biohydrogen, bioelectricity and bioproducts, will be affected by production decisions, conversion route, regionality, etc. The uncertainties and constraints related to a steady and sustainable feedstock supply construes another major limitation, especially under a changing climate (Mai et al., 2022).

Even as bioenergy pathways are understood as critical to achieving low-carbon energy and carbon dioxide removal (CDR) simultaneously, there is substantial ambiguity in evaluating these pathways. Different studies may use disparate data sources and assumptions to estimate costs. Hence, there is a need for consistent projections of prices and quantities for fuels and energy carriers at a sufficient level of detail that would help resolve data gaps and discrepancies in bioenergy systems' representation. Given the need for consistency in analyzing bioenergy systems and for harmonizing the evaluation of pathways for using biomass feedstocks to meet economy-wide decarbonization goals, this study aims to provide a harmonized supply curve of future fuels and energy carriers in the United States out to 2050. This could inform stakeholders on potential costs of alternative technology choices, key uncertainties, and future analysis to improve the understanding of the feasibility and tradeoffs of alternative decarbonization pathways. Furthermore, it could help feed into top-down models, thus being instrumental in soft-linking.

Methodology

We developed a harmonized data framework that brings together the input requirements for production of fuels and energy carriers and linking them with data for biomass supply, and price of the inputs (Figure 1). The input requirements for fuels and energy carriers were gathered from existing, Department of Energy (DOE)-sponsored process modeling and other peer-reviewed techno-economic analysis (TEA). The data compiled from different TEA sources were harmonized including itemized cost of inventories, dollar year for pathway cost estimations, cost of feedstock at reactor throat etc. For each of the subsections in figure 1 (electricity, hydrogen and biofuel), we employed different scenario narratives representing a sufficiently diverse set of conditions, including varying levels of technology advancement, ease of deployment, and supporting policy conditions. Under the developed set of scenarios, we analyzed the impact of regional variation in hydrogen and electricity retail rate on the price of biofuels using the Regional Energy Deployment System (ReEDS) model (Cohen et al., 2018), Hydrogen Analysis Production (H2A) model (NREL, 2024) and hydrogen delivery scenario analysis model (HDSAM) (ANL, 2024). Using this framework, we estimated supply curves for these fuels and energy carriers out to 2050 for the U.S. and the 5 U.S. Petroleum Administration Defense Districts (PADDs).

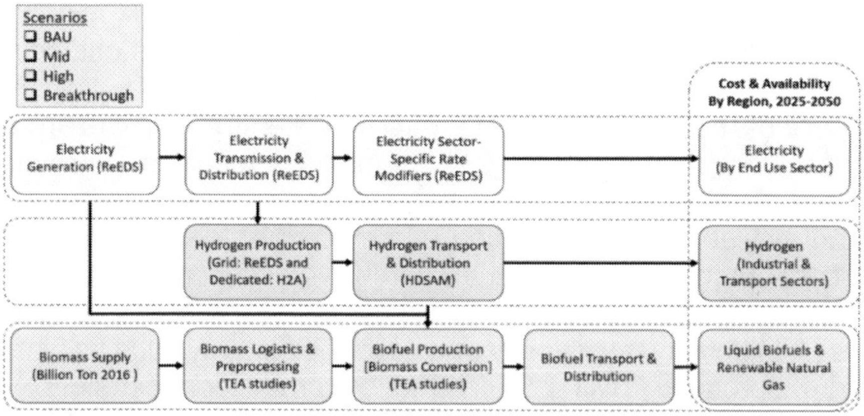

Figure 1. Analysis framework. Yellow boxes in the figure denote the approach for estimating electricity cost and availability, both electricity as end-use consumption and as an input to other fuel production processes. Blue boxes denote the approach for estimating hydrogen cost and availability, both as end-use consumption and as an input to other fuel production processes and green boxes for liquid biofuels.

Bioenergy Pathways Selection

This study covered different bioenergy pathways selected based on a set of clear criteria including technology readiness level, fuel yield, target fuel and carbon intensities of fuels. The selected pathways include biomass-to-biofuel, electricity, hydrogen, and biogas (Table 1). However, this paper predominantly focuses on biofuel pathways based on prior screening (Hawkins et al., 2024). This set of pathways was selected to represent potential options for the conversion of wet wastes, herbaceous and woody cellulosic feedstocks to different products. For most of these pathways, we incorporate varying levels of carbon capture and sequestration (CCS) to serve as a CDR strategy. For example, the conversion pathway for herbaceous biomass to jet fuel includes three variations: with no CCS, partial CCS (fermentation CO_2), and full CCS (fermentation and combustion CO_2).

Table 1. Biomass utilization pathways considered.

Feedstock Category	Conversion Process	Intermediates	Products	Reference
Herbaceous	Biochemical conversion + ethanol to jet (ETJ) (with different levels of CCS)	Ethanol	Jet Fuel (target) Gasoline	Hawkins et al., 2024
Woody	Gasification and Fischer-Tropsch synthesis (FT-SPK) (with different levels of CCS)	Syngas	Jet Fuel (target) Diesel Naphtha	Hawkins et al., 2024
Woody	Gasification and synthesis (with different levels of CCS)	Methanol	Methanol	Hawkins et al., 2024
Woody	Catalytic fast pyrolysis (CFP) and upgrading (with different levels of CCS)	Pyrolysis oil	Jet Fuel (target) Diesel	Hawkins et al., 2024
Herbaceous and Woody	Gasification to Bio-Hydrogen with CCS	Hydrogen	Hydrogen	Spath et al., 2005
Herbaceous and Woody	Bioelectricity with CCS	N/A	Electricity	Buchheit et al., 2021
Wet Waste	Hydrothermal liquefaction (HTL) and hydrotreating	Biocrude	Jet Fuel Diesel	Snowden-Swan et al., 2022
Wet Waste	Anaerobic digestion (AD)	Biogas	Natural Gas	Bhatt and Tao, 2020
Corn	Fermentation + ETJ	Ethanol	Jet Fuel (target) Gasoline	Tao et al., 2017
Fat, oil and greases (FOGs)	Hydrotreated Esters and Fatty Acids (HEFA)	N/A	Jet Fuel (target) Gasoline Diesel	Tao et al., 2017

Estimation of Pathway Minimum Fuel Selling Price (MFSP)

The production cost of hydrocarbon fuels, also known as minimum fuel selling price (MFSP) were estimated with the material and energy balances from the process models using a discounted cash flow rate of return analysis at a discount rate of 10% and net present value of zero (Davis et al. 2013, Dutta et al. 2011). The MFSP is reported in 2020 U.S. dollar and gallon gasoline equivalent basis (GGE).

Biomass Availability

The supply of biomass for producing liquid biofuels, biohydrogen, electricity and other bioenergy products was aligned with the scenarios of agricultural productivity increases in the 2016 Billion Ton Report (BT16) (Langholtz et al., 2016). Scenario differentiation followed the different assumptions of biotechnology-driven improvements in energy crop productivity potential explored in BT16. We assume biomass availability from BT16 based on 1, 2, 3, and 4% annual improvement in yields for the BAU, Mid, High, and Breakthrough scenarios. Figure 2 shows the supply curve of biomass under different scenarios we considered. The supply curve shows biomass price at the reactor throat (farmgate price plus the logistic and pre-processing cost) and availability assuming 30% processing losses and stranded biomass from raw biomass at the farmgate to finished feedstock at the reactor throat (Hawkins et al., 2024). As the price of biomass increases, the supply increases and 1 billion tons of biomass can be made available at $132/dry ton under the breakthrough scenario. There is significant potential for sustainably-produced biomass in the U.S. with supplies ranging from 0.7–1.1 billion tons across the scenarios analyzed. These values are in addition to ~300 million tons biomass already being produced and used in the U.S.

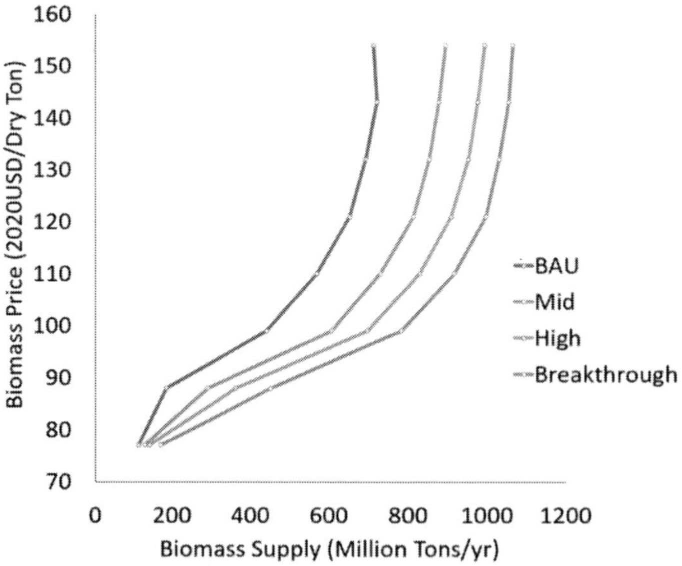

Figure 2. Biomass supply curve based on BT16 1, 2, 3 and 4% annual improvement in yields.

Key Results And Findings

Contributions to Bioenergy Costs

The price of biofuels production in this study ranged from ~\$3.00-7.00/GGE with biomass reactor throat price range of \$77/dry ton to \$154/dry ton for a facility processing 2,205 dry tons per day of biomass feedstock. The production costs for biofuel pathways without CCS start from ~\$3.00-\$5.00/GGE depending on the pathway. Figure 3 shows production cost at \$110/dry ton reactor throat price. Bioenergy costs are dominated by feedstock and capital costs, although these are not uniformly applied to all cases (for example in biogas production where other operating costs like utilities and cost of other purchased inputs dominate). Some pathways offer similar price points as fossil incumbents (e.g., FT-SPK for SAF or even methanol that can replace diesel). Biohydrogen is particularly quite cost-effective when compared to electrolysis at current technology (Hubert et al., 2024). For most biomass conversion pathways considered, there are opportunities for incorporating CCS due to availability of multiple flue gas streams. For some pathways, the cost of CCS is incremental (≤12%) relative to the case with no CCS. However, in some cases, flue gas streams have low CO_2 purity, that make CCS costlier and increase the MFSP by up to 45% relative to the case with no CCS.

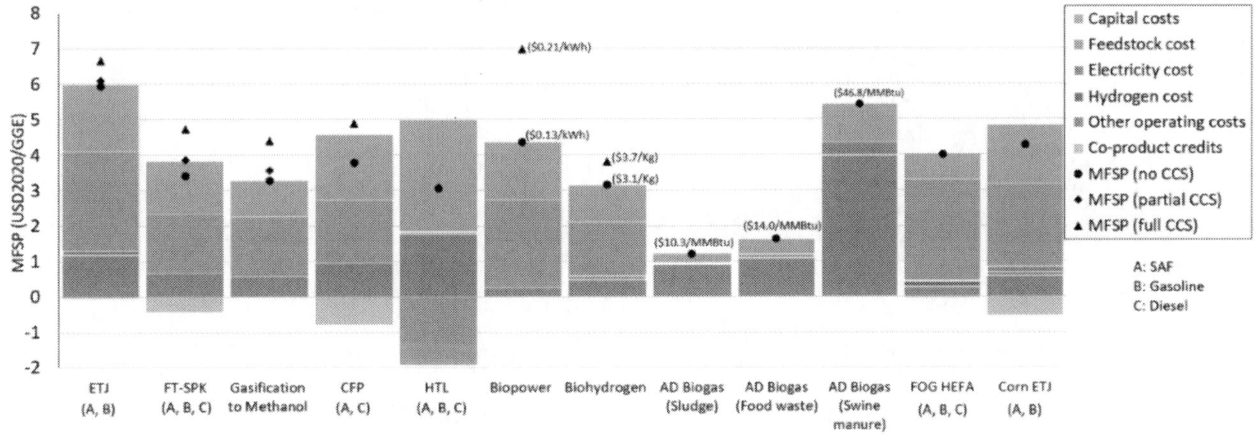

Figure 3. MFSP of bioenergy pathways considered (costs for U.S. average in 2050 at fixed feedstock costs of 110 USD/dry ton reactor throat price, Sludge at -185 USD/dry ton, FOGs at 683 USD/dry ton and Corn at 166 USD/dry ton).

Supply Curves for Biofuels

Biofuel production potential in this study ranged from 45-65 BGGE in 2050. Figure 4a shows the cost and availability estimate for biofuels by feedstock in 2050 if all biomass is used to produce biofuels. The stacked area shows the mid-scenario. Herbaceous biomass is the largest share of biomass potential. Hence, accelerating herbaceous biomass production from crop residues and dedicated energy crops will be critical to unlocking biofuel production. Although herbaceous is the highest overall, it only comes into play at ~\$5/GGE. So, the low-hanging fruits are the woody biomass, municipal solid wastes, and wet wastes. However, their availability is limited; herbaceous is needed for reaching the coveted billion-ton mark.

In addition, figure 4b shows the supply curve for biofuels in 2050. The biofuel supply curves are aggregate of fuel supply from multiple pathways and feedstocks at a specific price. The figure includes biofuels from cellulosic ethanol to jet, catalytic fast pyrolysis, gasification and FT-synthesis, gasification to methanol, hydrothermal liquefaction of wet wastes, HEFA and corn ethanol to jet. Biofuel volumes increase steeply with price, following biomass supply curve (Figure 2). The results show that a significant volume of biofuels could be produced at reasonable prices with a potential maximum supply between 45-65 BGGE. Waste-based fuels are available in smaller quantities at lower costs. Woody feedstocks then come online at an intermediate production cost and moderate volumes. The largest share of fuel is produced from herbaceous cellulosic feedstocks, through ethanol-jet pathway and have a higher production cost.

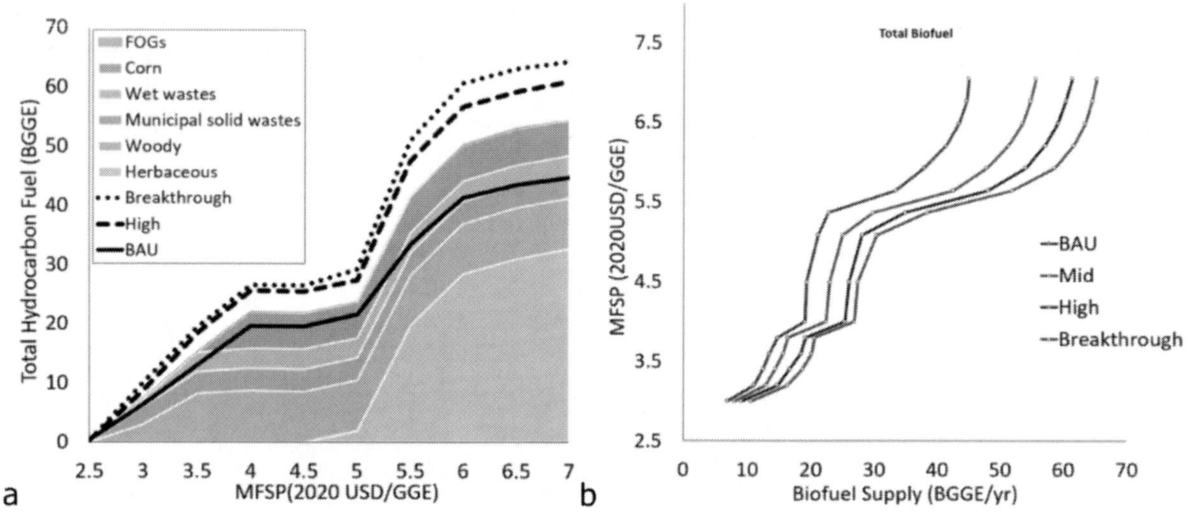

Figure 4. (a) Biofuel production by feedstock categories, (b) Overall biofuel supply curve (potential if all biomass is used to produce biofuels).

Conclusions

Scaling up emerging technologies will be critical to tackling the most pressing climate challenges and achieving net zero emissions by 2050. The result from this study showed that a substantial amount of bioenergy can be produced at costs comparable to fossil incumbents; however, upfront investment will be required for scale up. Biofuels production potential ranged between 45-65 billion gasoline equivalent (GGE) in 2050 and production costs start from ~\$3.00-\$5.00/GGE depending on the pathway. The estimated biomass supply is sufficient to provide 35 billion gallons of sustainable aviation fuel (SAF) and 60 billion gallons of total fuel (including other biorefinery coproduct fractions) in the "high" technology ambition scenario at production costs of \$7/GGE. Although the results here place emphasis on biofuels, there are other energy carriers (bioelectricity, biohydrogen) and products that can be produced from biomass depending on incentives and decarbonization content. The cost and availability of each of these bioenergy categories will depend on decisions to produce, convert, and use biofuels, bioelectricity, biohydrogen and bioproducts given the potential competition for the same feedstock. The results from this study help draw insights about the interplay of feedstock availability, technology readiness and potential costs of alternative technology choices.

Acknowledgement

This work was partially performed as part of the Decarbonizing Energy through Collaborative Analysis of Routes and Benefits (DECARB) program and the authors gratefully acknowledge the financial support provided by the U.S. Department of Energy (DOE) Office of Energy Efficiency and Renewable Energy (EERE). The views expressed in the article do not necessarily represent the views of the DOE or the U.S. Government. The U.S. Government retains and the publisher, by accepting the article for publication, acknowledges that the U.S. Government retains a nonexclusive, paid-up, irrevocable, worldwide license to publish or reproduce the published form of this work, or allow others to do so, for U.S. Government purposes.

References

1. U.S. Department of State and U.S. Executive Office of the President (2021), "The Long-Term Strategy of the United States: Pathways to Net-Zero Greenhouse Gas Emissions by 2050," https://www.whitehouse.gov/wp-content/uploads/2021/10/US-Long-Term-Strategy.pdf (Accessed November 10,2024).
2. Mai, Trieu, Denholm, Paul, Brown, Patrick, Cole, Wesley, Hale, Elaine, Lamers, Patrick, Murphy, Caitlin, Ruth, Mark, Sergi, Brian, Steinberg, Daniel, and Samuel, F. Baldwin (2022), "Getting to 100%: Six Strategies for the Challenging Last 10%," Joule, pp. 1981-1994.
3. Stuart, Cohen, Becker, Jon, Bielen, Dave, Brown, Maxwell, Cole, Wesley, Eurek, Kelly, Frazier, Will, Frew, Bethany, Gagnon, Pieter, Ho, Jonathan, Jadun, Paige, Mai, Trieu, Mowers, Matthew, Murphy, Caitlin, Reimers, Andrew, Richards, James, Ryan, Nicole, Spyrou, Evangelia, Steinberg, Daniel, Sun, Yinong, Vincent, Nina, and Zwerling, Matthew (2019), "Regional Energy Deployment System (ReEDS) Model Documentation: Version 2018," National Renewable Energy Laboratory, Technical Report NREL/TP-6A20-72023, pp.1-124.
4. National Renewable Energy Laboratory (2024), "H2A: Hydrogen Analysis Production Models," https://www.nrel.gov/hydrogen/h2a-production-models.html (accessed November 5, 2024).

5. Argonne National Laboratory (2024), "Hydrogen Delivery Scenario Analysis Model (HDSAM)," https://hdsam.es.anl.gov/index.php?content=hdsam (accessed November 5, 2024).
6. Hawkins Troy R., Tao, Ling, Binsted, Matthew, Burli, Pralhad, Field, John, Singh, Udayan, Horowitz, Russell, Oke, Doris, Bhatt, Arpit, Hartley, Damon, Kar, Saurajyoti, Lin, Yingqian, Paudel, Rajiv, and Ookie Ma (2024), "The Role of Biofuels and Biomass Feedstocks for Decarbonizing the U.S. Economy by 2050," National Renewable Energy Laboratory, PP.1-148.
7. Spath, P. Aden, A. Eggeman, T. Ringer, M. Wallace, B. and Jechura, J. (2005), "Biomass to Hydrogen Technical Report and Economics Utilizing the Battelle Columbus Laboratory Indirectly Heated Gasifier," Technical Report NREL/TP-510-37408, National Renewable Energy Laboratory, PP.1-57
8. Buchheit, Kyle, Lewis, Eric, Mahbubani, Kishore and Carlson, Derrick (2021),"Technoeconomic and Life Cycle Analysis of Bio-Energy with Carbon Capture and Storage (BECCS) Baseline," National Energy Technology Laboratory, PP. 1-278.
9. Snowden-Swan, Lesley, Li, Shuyun, Thorson, Michael, Schmidt, Andy, Cronin, Dylan, Zhu, Yunhua, Hart, Todd, Santosa, Miki, Fox, Sam, Lemmon, Teresa, and Swita, Marie (2022), "Wet Waste Hydrothermal Liquefaction and Biocrude Upgrading to Hydrocarbon Fuels: 2022 State of Technology," Pacific Northwest National Laboratory, pp. 1-22.
10. Bhatt, H. Arpit and Ling, Tao (2020), "Economic Perspectives of Biogas Production via Anaerobic Digestion,"Bioengineering, pp. 1-19.
11. Tao, Ling, Markham, N. Jennifer, Haq, Zia and Biddy, J. Mary (2017), "Techno-economic Analysis for Upgrading the Biomass-derived Ethanol-to-jet Blendstocks, Green Chemistry, pp.1082-1101.
12. Davis, R., Tao, L., Tan, E. C. D., Biddy, M. J., Beckham, G. T., Scarlata, C., Jacobson, J., Cafferty, K., Ross, J., Lukas, J., Knorr, D. and Schoen, P. (2013), "Process Design and Economics for the Conversion of Lignocellulosic Biomass to Hydrocarbons: Dilute-Acid and Enzymatic Deconstruction of Biomass to Sugars and Biological Conversion of Sugars to Hydrocarbons," National Renewable Energy Laboratory, NREL/TP-5100-60223, pp.1-139.
13. Dutta, A., Talmadge, M., Hensley, J., Worley, M., Dudgeon, D., Barton, D., Groenendijk, P., Ferrari, D., Stears, B., Searcy, E. M., Wright, C. T. and Hess, J. R. (2011), "Process Design and Economics for Conversion of Lignocellulosic Biomass to Ethanol," NREL/TP-5100-51400, pp. 1-176.
14. Langholtz, M. H., Stokes, B. J. and Eaton, L. M. "2016 Billion-Ton Report: Advancing Domestic Resources for a Thriving Bioeconomy," Oak Ridge National Laboratory, ORNL/TM-2016/160, pp. 1-411.
15. Hubert, McKenzie, Peterson, David, Miller, Eric, Vickers, James, Mow, Rachel and Howe, Campbell (2024), "Clean Hydrogen Production Cost Scenarios with PEM Electrolyzer Technology" https://www.hydrogen.energy.gov/docs/hydrogenprogramlibraries/pdfs/24005-clean-hydrogen-production-cost-pem-electrolyzer.pdf?sfvrsn=8cb10889_1 (Accessed November, 10 2024).

134b Sustainability Assessment of Grass to Sustainable Fuel Pathways

Pallavi Dubey, Lisa Schulte, Mark Mba Wright

The National Renewable Energy Laboratory has pioneered a biorefining process capable of converting methane-forming wet waste into energy-dense hydrocarbons through anaerobic digestion [1]. In our recent study, we leverage this innovation to address reducing greenhouse gas (GHG) emissions within the aviation sector by converting mixed Prairie biomass to Jet fuel. This project is supported by a $80 million USDA Climate Smart Award to demonstrate prairie grass to renewable fuel technologies at the commercial scale. Our approach involves a comprehensive Techno-Economic Analysis (TEA) of utilizing anaerobic digestion (AD) for Sustainable Aviation Fuel (SAF) production, encompassing a multi-step process spanning feedstock selection, biogas production, refinement, and catalytic conversion of biomethane into hydrocarbons, all culminating in the production of aviation-grade fuels [2]. Expanding upon our prior investigations [2], this study advances the analysis by broadening the scope of our Life Cycle Assessment (LCA) to include an array of environmental factors. While our previous research established the minimum fuel selling price in range $30-40 of AD-derived SAF and highlighted a promising Global Warming Potential (GWP) of -0.14 Kg CO2eq/MJ due to greenhouse gas capture, our current work delves deeper into the ecological ramifications of the AD-SAF pathway. This follow-up analysis expands our environmental assessment by integrating a wider range of impact indicators. We now consider factors like acidification, ecotoxicity, eutrophication, and human health effects alongside global warming and air quality concerns. This comprehensive evaluation aims to provide a holistic understanding of the ecological sustainability of AD-SAF production. This knowledge contributes to a more informed discussion about the potential of AD-SAF in fostering a sustainable future for aviation, specifically focusing on the environmental implications of using prairie biomass.

134c Modeling and Analysis of Sustainable Aviation Fuel Production Pathways

Pratyush Agarwal, Zhen Hou, Lili Yu, Lingxiang Li, Pushpesh Sharma, Darin Campbell, Shu Wang

As the aviation industry confronts the pressing need to mitigate its environmental footprint, pathways for the production of sustainable aviation fuel (SAF) have gained paramount importance. Currently, there are 7 broad categories of SAF production specified in ASTM D7566 Annexes: A1) Fischer-Tropsch synthetic paraffinic kerosene (FT-SPK), A2) hydroprocessed esters and fatty acids (HEFA-SPK), A3) hydroprocessed fermented sugars synthetic iso-paraffins (HFS-SIP), A4) FT-SPK with aromatics (FT-SPK/A), A5) alcohol-to-jet (ATJ-SPK), A6) catalytic hydrothermolysis jet (CHJ), and A7) hydrocarbon-hydroprocessed esters and fatty acids (HC-HEFA-SPK). This presentation aims to provide a comprehensive understanding of the technological approaches and environmental implications for the ASTM-approved pathways to produce SAF via models in Aspen Plus and HYSYS. The documentation and simulation files for the models are available on the Aspentech website.[1] Each of the pathways can be modeled with detailed reaction, thermodynamic, and process considerations to set up material and energy balances needed to determine the economic and environmental costs. Reduced-order hybrid representations of these detailed models can serve as options in a superstructure optimization problem, where these costs, along with blending limitations and governmental mandates, can help drive sustainability decisions in the aviation sector.

[1] Aspentech Sustainability Library: https://www.aspentech.com/en/insights/models/sustainability-models

134d Comparing Ways to Incorporate Biomass into E-Fuel Production

Simone Mucci, Alexander Mitsos, Dominik Bongartz

E-fuels produced with renewable electricity as main energy input are receiving attention as sustainable alternative to fossil fuels, with multiple potential target fuels being discussed [1]. While the first step in e-fuel production is usually water electrolysis to produce hydrogen [2], the production of organic fuels requires a carbon source. The choice of carbon source can have a noticeable impact on the energetic, economic, and environmental performance of the production process. The most common carbon source considered for e-fuel production is carbon dioxide, which can be captured from various point sources or air [3]. However, another promising carbon source is biomass, which will also be available in a low-fossil future supply system, and – unlike carbon dioxide – also constitutes an additional energy input into the system. Conversely, incorporating hydrogen from renewable electricity has been proposed as a way to improve the carbon efficiency and economics of biofuel production pathways [4,5].

There is a variety of ways in which biomass and renewable electricity can be combined to produce a given e-fuel. In particular, biomass can be converted to an intermediate stream suitable for the synthesis of downstream products via different conversion processes. The most versatile of these in terms of allowable feedstock are thermochemical conversion [6], i.e., either combustion (combined with an energy recovery process for utilizing the generated heat) or gasification to produce streams rich in carbon dioxide or carbon monoxide, which can then be combined with green hydrogen for e-fuel production. Similarly, for most downstream products, there are multiple options for producing them from hydrogen, carbon monoxide, and carbon dioxide. For example, synthetic kerosene can be produced via the Fischer-Tropsch pathway or the methanol-to-olefins pathway [7], each of which could be based either on biomass gasification or combustion. In this contribution, we thus compare pathways for producing synthetic kerosene from renewable electricity and biomass in terms of efficiency and economics and discuss inherent thermodynamic differences via exergy analysis. We find that gasification-based pathways benefit (i) from lower exergy destruction in the biomass conversion step than combustion-based pathways, and (ii) from the lower hydrogen demand that requires smaller water electrolyzers, which (at least in case of the more mature low-temperature electrolyzers) are the main contributors to exergy destruction. The latter also has a significant impact on production cost. Furthermore, the methanol-to-olefins pathway is found to have advantages over Fischer-Tropsch pathway for producing kerosene regardless of the biomass conversion technology.

Acknowledgments

The authors gratefully acknowledge the financial support by the German Federal Ministry of Education and Research (BMBF) within the H2Giga project DERIEL (grant number 03HY122D).

References

[1] Bongartz, L. Doré, K. Eichler, T. Grube, B. Heuser, L.E. Hombach, M. Robinius, S. Pischinger, D. Stolten, G. Walther, and A. Mitsos. Comparison of light-duty transportation fuels produced from renewable hydrogen and green carbon dioxide. *Applied Energy*, 231:757-767, 2018.

[2] Mucci, A. Mitsos, and D. Bongartz. Power-to-X processes based on PEM water electrolyzers: A review of process integration and flexible operation. *Computers & Chemical Engineering*, 175:108260, 2023.

[3] Von Der Assen, L.J. Müller, A. Steingrube, P. Voll, and A. Bardow. Selecting CO_2 sources for CO_2 utilization by environmental-merit-order curves. *Environmental Science & Technology*, 50:1093-1101, 2016.

[4] Hillestad, M. Ostadi, G. Alamo Serrano, E. Rytter, B. Austbø, J. Pharoah, and O. S. Burheim. Improving carbon efficiency and profitability of the biomass to liquid process with hydrogen from renewable power. *Fuel*, 234:1431–1451, 2018.

[5] König, K. Ulonska, A. Mitsos, and J. Viell. Optimal applications and combinations of renewable fuel production from biomass and electricity. *Energy and Fuels*, 33:1659–1672, 2019.

[6] Shahbaz, A. AlNouss, I. Ghiat, G. Mckay, H. Mackey, S. Elkhalifa, and T. Al-Ansari. A comprehensive review of biomass based thermochemical conversion technologies integrated with CO_2 capture and utilisation within BECCS networks. *Resources, Conservation and Recycling*, 173:105734, 2021.

[7] Schmidt, V. Batteiger, A. Roth, W. Weindorf, and T. Raksha. Power-to-liquids as renewable fuel option for aviation: A review. *Chemie-Ingenieur-Technik*, 90: 127–140, 2018.

HeatPath^{TM} catalyst technologies and DETCHEM^{TM} accessible simulation tools to produce Sustainable Aviation Fuels

Anna Lee Tonkovich, Tonkomo LLC, Gilbert, AZ
Eric Daymo, Tonkomo LLC, Gilbert, AZ
Yong Wang, Washington State University, Pullman, WA
Cody Lockhart, HeatPath Solutions/Nexceris, Lewis Center, OH
Matt Seabaugh, HeatPath Solutions/Nexceris, Lewis Center, OH
Akash Shirsath, Karlsruhe Institute of Technology, Karlsruhe, Germany
Matthias Hettel, Karlsruhe Institute of Technology, Karlsruhe, Germany
Olaf Deutschmann, Karlsruhe Institute of Technology, Karlsruhe, Germany

Abstract

HeatPath^{TM} is a catalyst carrier with a high thermal conductivity SiC-Al_2O_3 composite core, bonded to a hermetic and roughened protective alumina surface layer. HeatPath's enhances bulk tablet thermal conductivity by 4-10X over conventional tableted catalysts (depending on catalyst formulation and reactor operating temperature), enabling rapid thermal response in highly thermic reactions. Also, the thin catalyst coating over the hermetic core eliminates internal catalyst sites that can promote byproduct formation or product decomposition. The application of HeatPath^{TM} catalysts to the two main process steps (syngas production and Fischer-Tropsch Synthesis, FTS) to produce Sustainable Aviation Fuel (SAF) is discussed. For Dry Reforming of Methane (DRM) at a steam to carbon ratio (S:C) down to 0.3, the HeatPath^{TM} catalyst ran for 450 hrs without coke formation on the exterior or interior of the catalyst, confirmed by SEM EDS of selected pellets. A conventional DRM catalyst pellet, meanwhile, deactivated when operated with a S:C below 1.5. Although no coke was found on the exterior of the conventional catalyst, coke formed in the interior of the catalyst. Analysis with the software DETCHEM^{TM} PBR identified a CO concentrating effect inside the catalyst pellet which leads to the carbon activity exceeding a value of one. When carbon activity exceeds a value of one, coke is thermodynamically favored to form. Meanwhile, for exothermic FTS, HeatPath^{TM} avoids localized hotspots that cause thermal runaway, allowing unprecedented operating temperature and reactor efficiency (high CO conversion and low CH_4 production) with stable long-chain FTS production. Even at high temperatures (>250°C) for the cobalt-based catalyst, HeatPath^{TM} maintains high fractions of high-carbon products. A kinetic model was fit with DETCHEM^{TM} PBR, demonstrating the versatility of this software for rapidly evaluating reactor performance.

Introduction

To meet the goals of the 2021 White House SAF Grand Challenge [1] the US must produce 3B gpy of SAF with >50% lower GHG emissions versus Jet A in 2030; and 35B gpy by 2050. <u>Global</u> 2022 SAF production was 26.5M gpy [2], less than 1% of the 2030 target. Current SAF routes cannot compete with Jet A on cost, but the Grand SAF Challenge increased airlines' financial incentive, adding $1.25/gal in SAF tax credits. As a result, there is intense interest in low-cost, low-carbon intensity SAF, particularly CO_2 neutral (or negative) blend-stocks. Achieving these aggressive goals requires many new technologies, including catalyst and simulation capabilities. In this work, we focus on two important chemistries for SAF (Fig. 1): syngas production from biogas by Dry Reforming of Methane (DRM) and liquid fuel production by Fischer-Tropsch Synthesis (FTS). In both cases, HeatPath™ catalysts were used to achieve superior performance compared to conventional catalysts.

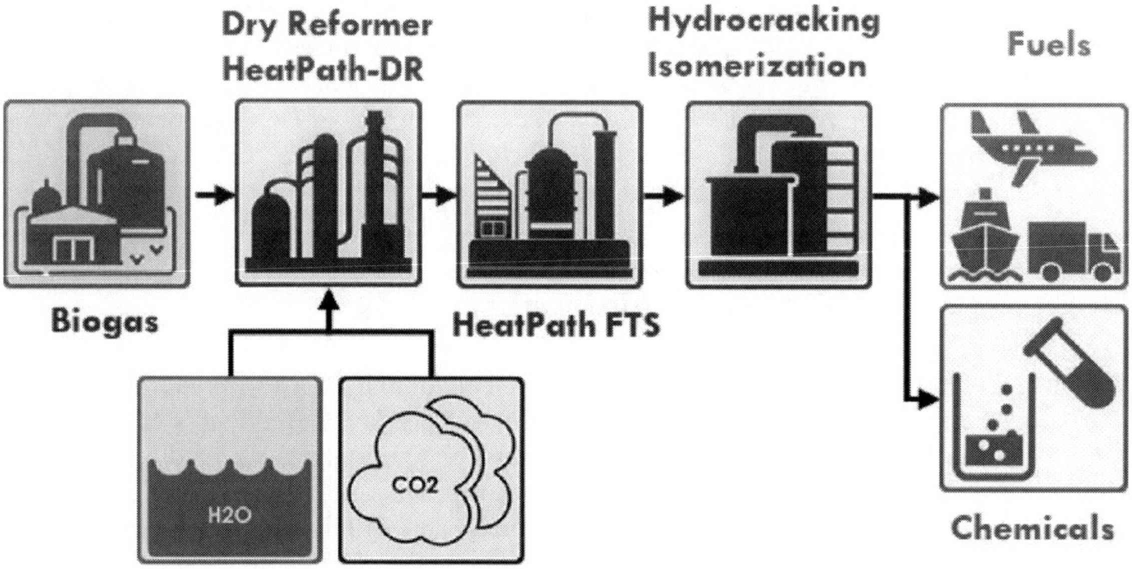

Figure 1. Process block diagram for SAF production from green hydrogen and captured CO_2.

HeatPath™ [3] is catalyst carrier with a high thermal conductivity SiC-Al_2O_3 composite core, bonded to a hermetic and roughened protective alumina surface layer (Fig 2). The rugged surface enhances catalyst coating adhesion, allowing the application of various catalysts. HeatPath's key innovations include the enhancement of bulk tablet thermal conductivity by 4-10X over conventional tableted catalysts (depending on catalyst formulation and reactor operating temperature), enabling rapid thermal response in highly thermic reactions. Further, the thin catalyst coating over the hermetic core eliminates internal catalyst sites that can promote byproduct formation or product decomposition.

Figure 2. Uncoated (left) and coated (right) HeatPath™ catalyst tablets.

With respect to intra-particle diffusion, practical catalysts balance non-linear multi-step reactions with second-order diffusion effects. This interplay can reveal unexpected findings that mask laboratory observations or scale-up problems. Critical diffusion dimensions that affect side reactions depend on the reaction network complexity. These critical diffusion lengths that inhibit side reactions may be on the order of just a few μm, or as large as 1000+ μm. The "secret life of catalysts" occurs inside catalyst structures (e.g., *pellets, powders, and wall coated catalysts*), where transport is controlled by diffusion and where series or series-parallel reactions can dominate. Although this internal region physically contains most catalyst sites, complex reactions inside pore structures are generally poorly analyzed and rarely optimized. Side reactions that run amok inside a catalyst can be conflated with an intrinsic catalyst problem and the catalyst practitioner may move to the next formulation or condition. Reduced performance at scale can become accepted, instead of being understood and improved upon.

The physics of coupled reaction and diffusion is not new, described by (among others) Turing [4], Aris [5] and Fogler [6]. Every undergraduate chemical engineer learns the principles of internal diffusion limitations and a simple effectiveness factor analysis in their reaction engineering class.

The problem of missing important chemistry occurring inside catalyst pellets and coatings lies in the use of simplified analyses. Reality is often more complex and interesting than a simplified analysis might conclude. So why is this known effect of internal diffusional resistance coupled with reaction mostly ignored or oversimplified? Accessible reaction engineering tools are not routinely available nor are results generally expected to challenge the interpretation of laboratory data. Here, accessible means readily usable with little programming or customization needed for the chemical engineer more interested in the application than the theory. Without such reactor simulation tools, it is difficult to see what one doesn't expect to find.

DETCHEM™ is a suite of accessible reaction engineering tools designed to help to understand what happens inside a catalyst, integrating:

- multi-step reaction kinetics, either from microkinetic models with perhaps hundreds of steps or apparent rates with a few steps, coupled with
- catalyst internal pore diffusion that may be molecular, Knudsen, or mixed, coupled with
- realistic heat transfer models, such as state-of-the art correlations for packed beds, structures, or open channels, to model the flow of heat that results from exterior heating/cooling along with endo- or exothermic heat transfer demands linked with the kinetic mechanisms.

DETCHEM™ PBR, a one-dimensional packed bed reactor simulator that is part of the DETCHEM™ software suite, is used for the present work to analyze reactions inside the pellets for DRM, and fit kinetics for FTS.

In Section 2, we investigate a commercial Dry Reforming of Methane (DRM) Nickel catalyst that was shown to concentrate CO inside the pellet, showcase how DETCHEM™ PBR is used to understand this effect, and demonstrate HeatPath™ as a solution. In Section 3 we showcase the superior FTS performance with HeatPath™ along with the use of DETHCEM PBR to rapidly fit kinetics for a global reaction mechanism.

Dry Reforming of Methane (DRM)

Summary experimental data for DRM collected with 6 mm Ni/alumina catalyst pellets are shown in Fig. 3A. The catalyst deactivated when operated with a S:C below 1.5 and a high pressure of 20 barg but there was no visible external coke. DETCHEM™ PBR, however, predicted the formation of coke in the interior of the catalyst pellet for most of the catalyst bed (Fig. 3B). This was confirmed by SEM EDS (Fig. 4). Coke forms in the interior of the catalyst due to the concentrating effect of CO inside the catalyst pellet (Fig. 5) which leads to the carbon activity exceeding one. When carbon activity exceeds a value of one, coke is thermodynamically favored to form. DRM modeling was based on an experimentally fit apparent kinetics for a simplified reaction network consisting of DRM, Water Gas Shift (WGS), and Steam Methane Reforming (SMR) steps.

One solution to avoid coke formation in DRM reactors is to use the Nexceris HeatPath™ catalyst, whereby the catalyst is coated on a dense, highly conductive core (Fig. 6). At a steam to carbon ratio (S:C) down to 0.3 the catalyst ran for 450 hrs without coke formation (Fig. 7), as predicted by DETCHEM™ PBR modeling and confirmed by SEM EDS of selected pellet (Fig. 8). Despite the lower volumetric catalyst loading of the 12wt.% Ni doped Al_2O_3-CaO-MgO material when applied as a thin coating on the HeatPath™ catalyst particle, the performance improved (Table 1) due to the elimination of side reactions inside the catalyst pellet. Without the proper accessible tools, such as DETCHEM™ PBR, finding an engineering solution would be difficult.

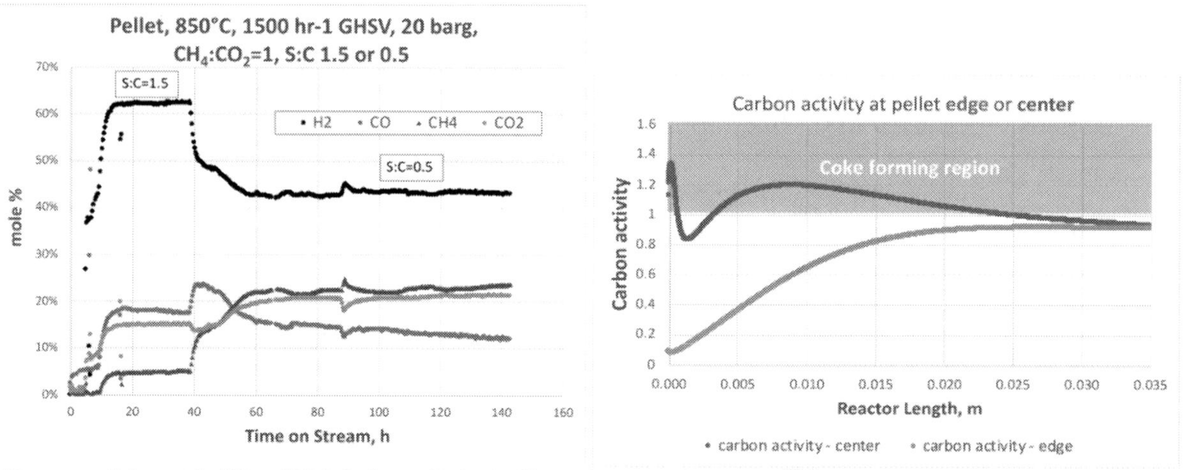

Figures 3A and 3B. DRM data (3A, left) and a DETCHEM™ PBR modeling prediction that coke under these experimental conditions should form inside the catalyst pellets, not on the exterior of the pellets (3B, right).

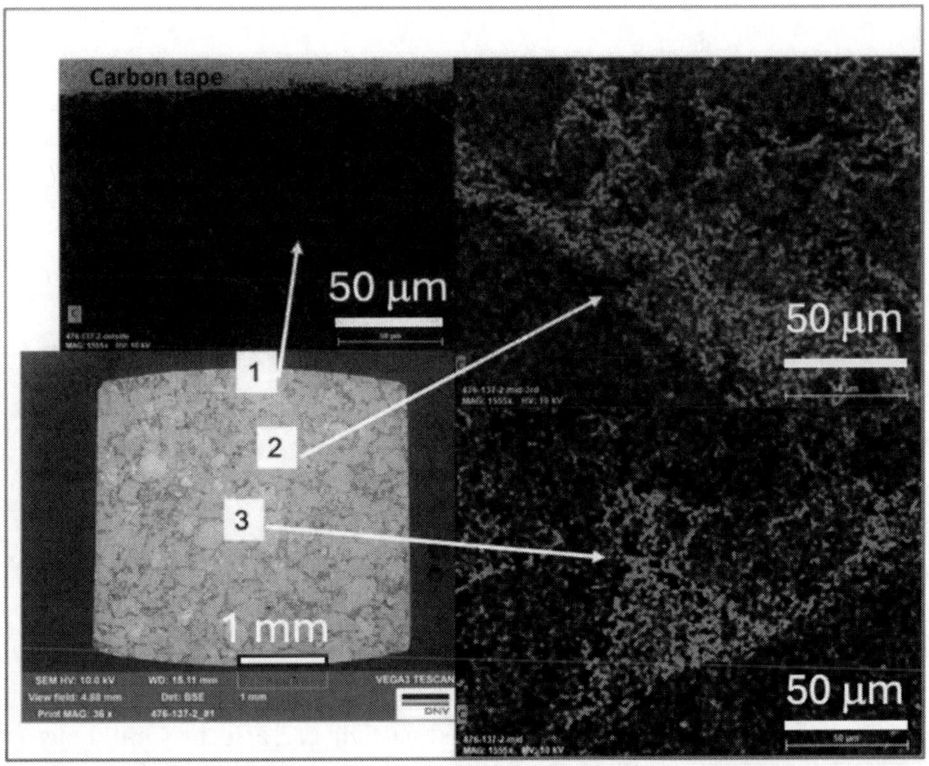

Figure 4. SEM analysis confirms the presence of coke inside, but not on the outside of the catalyst pellet, as predicted by DETCHEM™ PBR. The length scales have been enhanced for clarity.

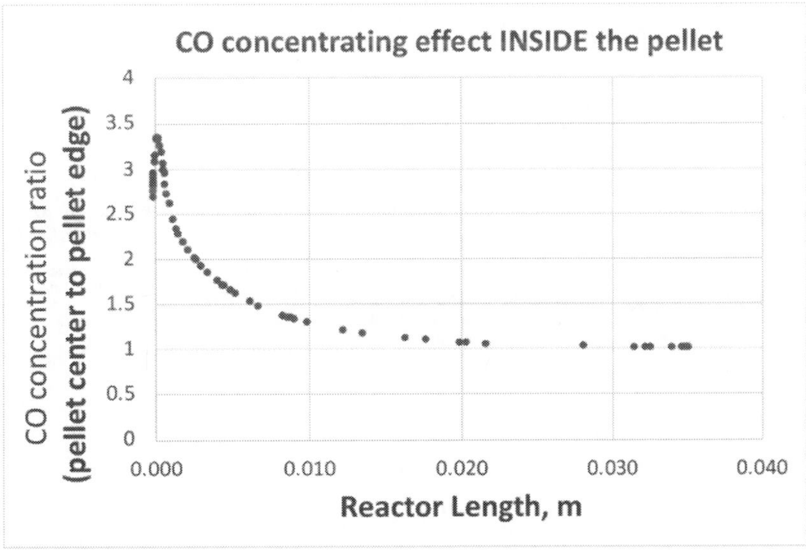

Figure 5. Coke forms due to a concentrating effect of CO inside the pellet, as predicted by DETCHEM™ PBR.

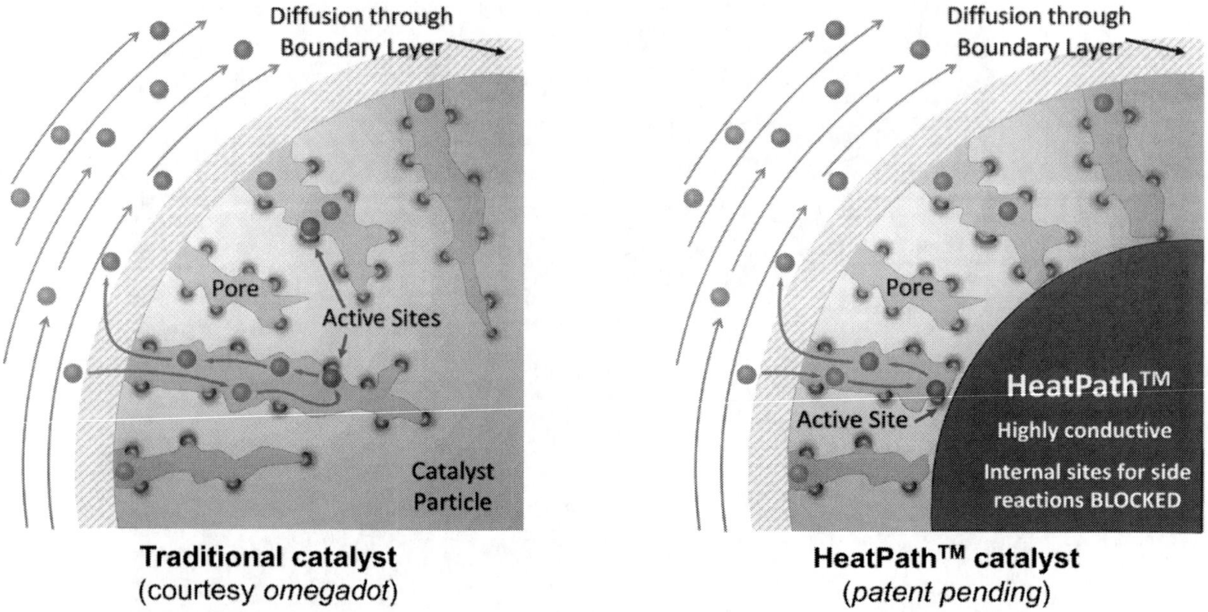

Figures 6. Conventional pellet catalysts (left) can have intricate pore structures that limit mass transfer. Mass transfer resistances inside a catalyst particle can affect overall reactor performance, but the effect of stagnant zones within the pellet can be mitigated by instead using a thin catalyst coating. Thin shell coatings on highly conductive catalyst cores are possible with Nexceris HeatPath™ (right).

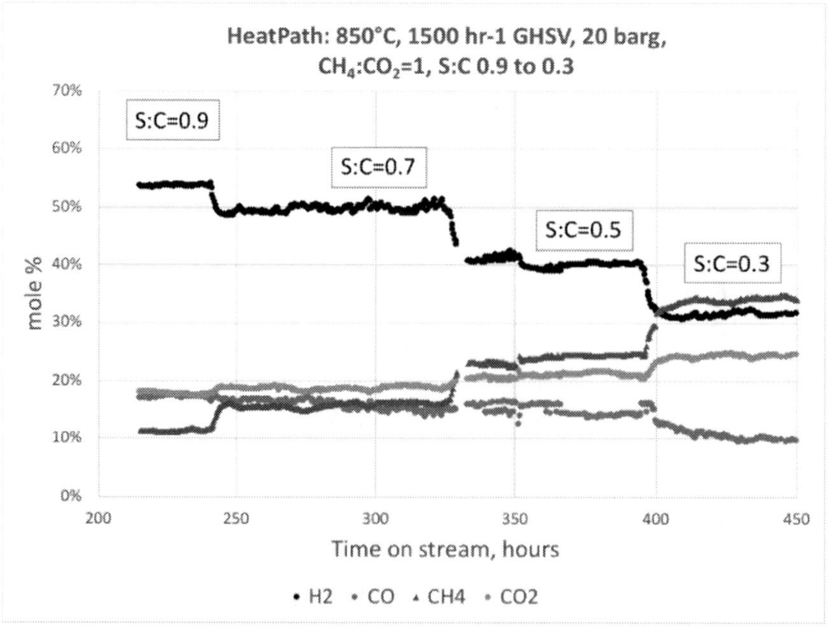

Figures 7. DRM tests were repeated with a HeatPath™ catalyst. Deactivation was not observed, even down to a steam-to-carbon ratio of 0.3.

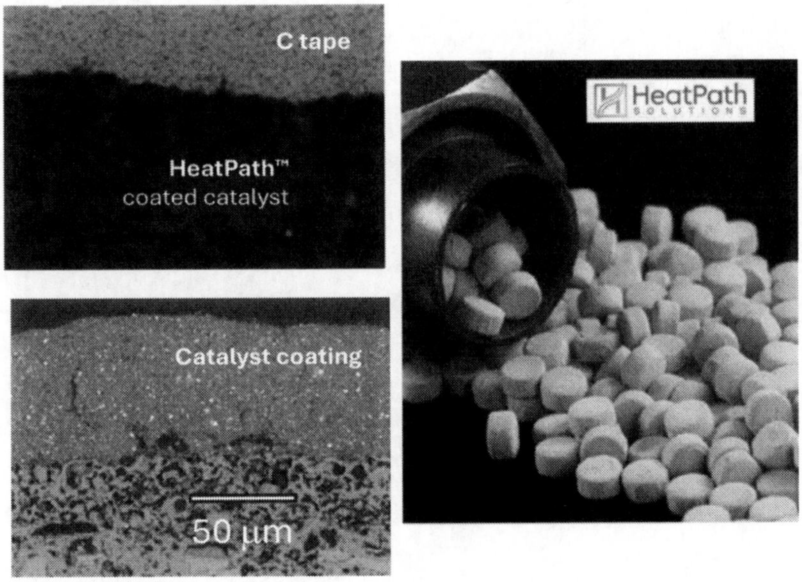

Figure 8. With the thin shell coating, coke does not form inside the catalyst (top left). The presence of carbon in the top of the SEM EDS dot map (top left, orange color) is due to carbon tape, not coke. The catalyst scale is shown on the bottom left, and unused catalyst is shown on the right.

Table 1. Summary of DRM results (GHSV=1500 hr^{-1}, P=20 barg, CH_4:CO_2=1, T=850°C). Despite the thin coating, the HeatPath™ pellet with a shell catalyst coating was more productive than a conventional pellet because the interior of the conventional pellet is where CO concentrated and unwanted side reactions – like coking – occurred.

System	S:C	g/g/h (CH_4)	g/g/h (CO_2)
Pellet	1.5	0.15	0.12
Pellet (*pre-coking*)	**0.5**	0.11	0.37
HeatPath™	1.5	5.88	4.38
HeatPath™	**0.5**	2.95	10.12

Fischer-Tropsch Synthesis (FTS)

While the DRM process demonstrated the benefits of HeatPath's thin coatings to mitigate unwanted side reactions (such as coking), FTS demonstrates the benefits of HeatPath's high thermal conductivity core to mitigate hot spots.

Higher temperature operation normally increases catalyst productivity due to faster reaction rates, but with FTS higher temperature operation traditionally leads to thermal runaway. In a comparison between FTS catalyst powders and HeatPath™ FTS catalyst tablets for 30 wt% Co % 4.5 wt% Re (Fig. 9), powders were stable in a 1 mm microchannel below 200°C, achieving ~2 g/g/h productivity [8]. Thermal runaway occurred with this powder at less than 200°C when tested in a

7 mm diameter tube. Meanwhile, the HeatPath™ FTS catalyst is stable up to 255°C with a productivity of 20 g/g/h.

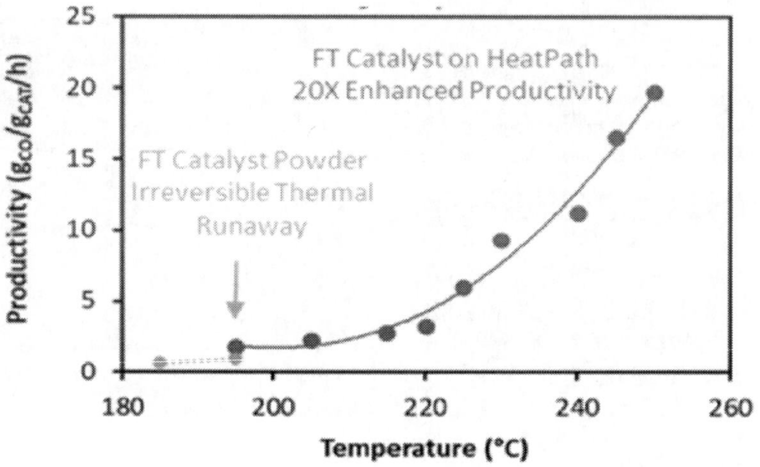

Figure 9. The high thermal conductivity of the HeatPath™ FTS catalyst led to a ~20X enhanced productivity, without thermal runaway, at 255°C. A conventional powdered FTS catalyst experienced runaway at less than 200°C.

In addition to operation at high temperature without thermal runaway, the HeatPath™ catalyst retained a high α (Flory-Schulz distribution) value of 0.806 at 255°C and a H_2:CO ratio of 1 (Figs. 10-11). Such unprecedented performance is likely due to the thin catalyst coating associated with HeatPath™ and is much improved over traditional FTS catalysts.

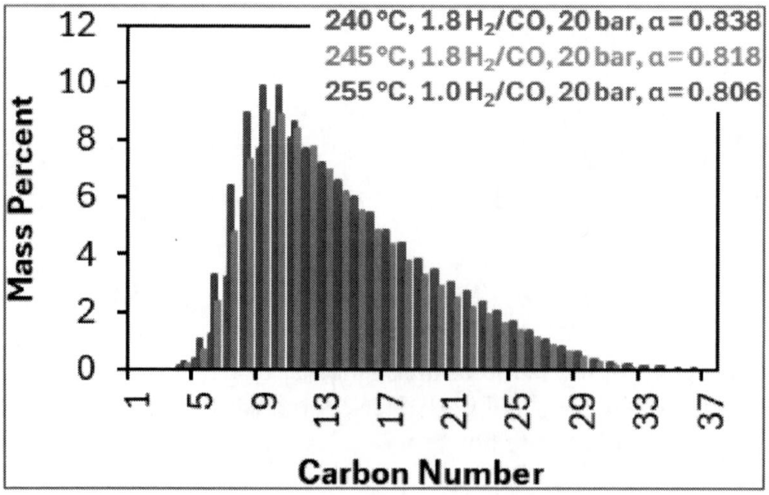

Figure 10. The product distribution produced under different operating conditions using the HeatPath™ FTS catalyst.

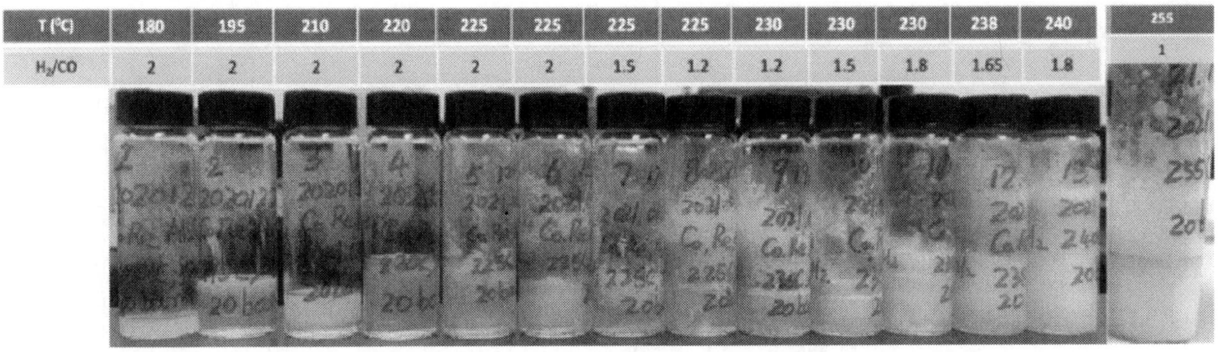

Figure 11. Imapges of the FTS products (the white material at the bottom is wax) produced with the HeatPath™ FTS catalyst (bottom). The associated temperature and H_2:CO ratio under which each sample was produced is shown above the sample images.

The collected catalyst test data were then used to fit kinetics for the formation of lights (CO_2, CH_4, C_2H_6, C_3H_8, and C_4H_{10}), along with the liquid/wax products lumped together as $C_{13}H_{28}$. The kinetics model, along with fitted activation energies and adsorption enthalpies, is summarized in Table 2. Fitting was accomplished by coupling the optimization toolkit DAKOTA [10] to DETCHEM™ PBR, and then using a genetic algorithm to find kinetics parameters that minimize the root mean squared difference between the data and the model for CO conversion and product selectivities. Around 100,000 cases were run find the optimized kinetics parameters, with each DETCHEM™ PBR run requiring just 1-2 core seconds. Such speed is only possible with accessible simulation tools like those available in the DETCHEM™ modeling suite. The resulting kinetics model fits well against the data (Fig. 12).

Table 1. Kinetics model and fit activation energy (Ea) and heat of adsorption (ΔH_{ads}) values for the HeatPath™ FTS catalyst. The rate constants, k_i, are given by $\boldsymbol{k_i = k_{0,i} \cdot e^{\left(\frac{-Ea}{RT}\right)}}$, and the adsorption constants, K_i, are calculated using $\boldsymbol{K_i = K_{0,i} \cdot e^{\left(\frac{\Delta H_i}{RT}\right)}}$.

Rxn	Lumped Stoichiometry	Rate Expression	Ea, kJ/mol	ΔH_ads, kJ/mol
1	$13\,CO + 27\,H_2 = C_{13}H_{28} + 13\,H_2O$	k1 Pco Ph2 / (1+K1 Pco)^2	83.6	-28.3
2	$CO + H_2O = CO_2 + H_2$	k2 Pco Ph2 / (1+K4 Pco)^2	68.7	31.8
3	$CO + 3\,H_2 = CH_4 + H_2O$	k3 Pco Ph2 / (1+K3 Pco)^2	119.1	37.9
4	$2\,CO + 5\,H_2 = C_2H_6 + 2\,H_2O$	k4 Pco Ph2 / (1+K4 Pco)^2	144.4	31.8
5	$3\,CO + 7\,H_2 = C_3H_8 + 3\,H_2O$	k5 Pco Ph2 / (1+K4 Pco)^2	142.8	31.8
6	$4\,CO + 9\,H_2 = C_4H_{10} + 4\,H_2O$	k6 Pco Ph2 / (1+K4 Pco)^2	139.4	31.8

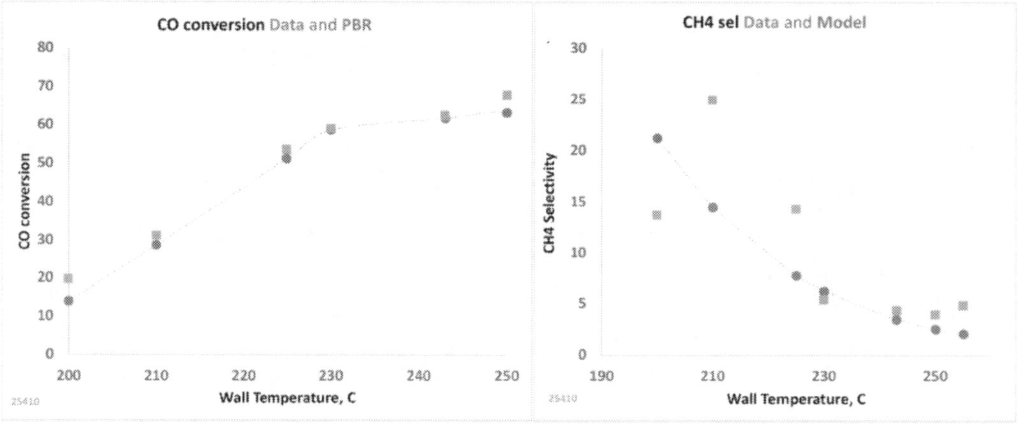

Figure 12. CO conversion and CH_4 selectivity are well predicted by the resulting kinetic fit using DAKOTA and DETCHEM™ PBR.

An assessment of the relative temperature control of the HeatPath™ catalyst is accomplished by analyzing the temperature dependence of the CO adsorption term, K_I, which appears in the denominator of the rate expression for $C_{13}H_{28}$ synthesis in Table 2. In Fig. 13, the temperature dependence of K_I is plotted for models of this parameter reported for various FTS kinetics models: HeatPath™ (Table 2), Yates & Satterfield [11], and Chabot [12]. [11] reported intrinsic kinetics collected with an isothermal batch reactor. The trend for K_I for HeatPath™ matches that of [11], suggesting that the HeatPath™ FTS catalyst demonstrated good thermal control. [12] tested 90 μm powders in a mini-channel reactor (2.75 mm gap thickness), with reported thermal control. As seen with the powdered catalyst testing with microchannels (Fig. 9), a mini-channel configuration may not be sufficient to ensure isothermal operation. Insufficient thermal control could be why K_I for [12] increases with temperature, while K_I decreases with temperature for HeatPath™ and [11], where good thermal control is maintained.

Conclusion

HeatPath™ and DETCHEM™ can accelerate the development of SAF processes.

HeatPath™ is a catalyst carrier that improves reactor performance. For DRM, HeatPath™ catalysts produce syngas without coking, while for FTS HeatPath™ catalysts form liquids at high productivity and low CH_4 selectivity, even at unprecedented high temperatures.

DETCHEM™ reaction engineering tools speed development by modeling diffusion limited processes inside catalysts and washcoats, fitting kinetics, augmenting experiments, and allowing for an understanding of the best operating window at lab and commercial scales. The speed, versatility, and ease-of-use make the DETCHEM™ suite well suited for the development of novel reactors and catalysts for decarbonized solutions.

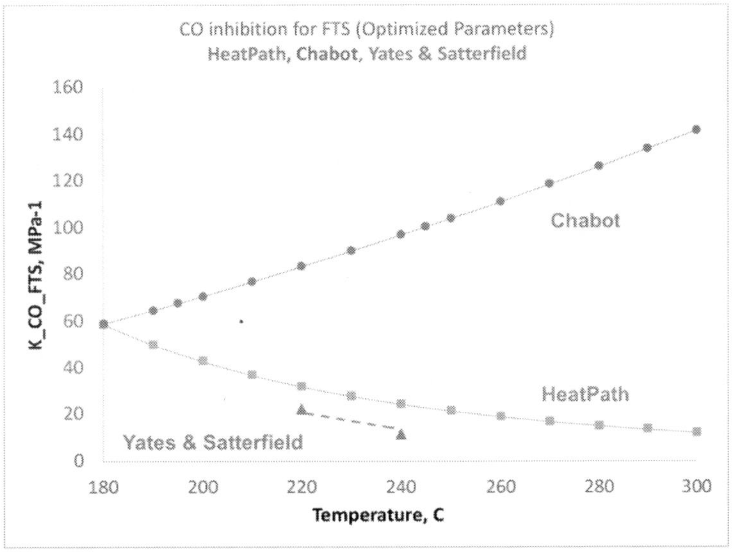

Figure 13. Assessment of the CO inhibition term (K_{CO}) suggests that the catalyst test reactor with HeatPath™ catalysts is under good thermal control. This is because the temperature dependence of K_{CO} matches the general trend reported by Yates & Satterfield [11], whose experiments had exceptional thermal control.

References

[1] DOE. 2021. "Memorandum of Understanding: Sustainable Aviation Fuel Grand Challenge." Sept. 8, 2021. https://www.energy.gov/sites/default/files/2021-09/S1-Signed-SAF-MOU-9-08-21_0.pdf . Accessed March 22, 2024.

[2] IATA - 2022 SAF Production Increases 200% - More Incentives Needed to Reach Net Zero. https://www.iata.org/en/pressroom/2022-releases/2022-12-07-01/ . Accessed March 22, 2024.

[3] HeatPath Solutions. https://heatpathsolutions.com . Accessed October 7, 2024.

[4] Turing A. The chemical basis of morphogenesis. *Phil. Trans. R. Soc. London* 1952; **237**: 37–72. DOI: https://doi.org/10.1098/rstb.1952.0012

[5] Aris R. The Mathematical Theory of Diffusion and Reaction in Permeable Catalysts: The theory of the steady state. Oxford Clarendon Press; 1975.

[6] Fogler HS. *Elements of Chemical Reaction Engineering*. New Jersey Prentice Hal; 1986.

[7] omegadot software & consulting, GmbH. DETCHEM™ v3.1 2024. www.omegadot.software . Accessed October 7, 2024.

[8] Cao C, Hu J, Li S., Wilcox W, Wang, Y. Intensified Fischer–Tropsch synthesis process with microchannel catalytic reactors. *Catal. Today* 2009; **140**: 149–156. https://doi.org/10.1016/j.cattod.2008.10.016

[9] Zhang R, Li J, Tonkovich A, Lockhart C, Wang X, Hu W, Karroum H, Seabaugh M, Kruse N, Wang Y. Highly productive and robust core@shell HeatPath SiC-Al_2O_3 @Co/Re/Al_2O_3 catalyst for Fischer–Tropsch synthesis. *App. Cat A* 2023; **666**: 119419. https://doi.org/10.1016/j.apcata.2023.119419

[10] Adams BM, Bohnhoff WJ, Dalbey KR, Ebeida MS, Eddy JP, Eldred MS, Hooper RW, Hough PD, Hu KT, Jakeman JD, Khalil M, Maupin KA, Monschke JA, Ridgway EM, Rushdi AA,

Seidl DT, Stephens JA, Swiler, LP, Winokur JG. *Dakota, A Multilevel Parallel Object-Oriented Framework for Design Optimization, Parameter Estimation, Uncertainty Quantification, and Sensitivity Analysis: Version 6.15 User's Manual* 2021. https://dakota.sandia.gov
[11] Yates IC, Satterfield CN. Intrinsic kinetics of the Fischer-Tropsch synthesis on a cobalt catalyst. *Energy & Fuels* 1991; **5**: 168-173. https://doi.org/10.1021/ef00025a029
[12] Chabot G, Guilet R, Cognet P, Gourdon C. A mathematical modeling of catalytic milli-fixed bed reactor for Fischer–Tropsch synthesis: Influence of tube diameter on Fischer Tropsch selectivity and thermal behavior. *Chemical Engineering Science* 2015; **127**: 72-83. https://doi.org/10.1016/j.ces.2015.01.015

134f Process Design for the Conversion of Lignocellulosic Biomass into Sustainable Aviation Fuel

Nurul Fuadi Pratiwi Sulaiman, Hyerim Kim, Kyeongsu Kim

Over the past century, global energy demand has followed an exponential growth trajectory. Key drivers include the expanding world population (now exceeding 7 billion) and rising welfare levels. Currently, the majority of our energy comes from fossil fuels such as crude oil, natural gas, and coal. However, the combination of high energy demand and limited fossil fuel supply has led to oil prices surpassing $100 per barrel. Unfortunately, heavy reliance on fossil fuels has also contributed to elevated CO_2 levels in the atmosphere. In recent decades, there has been growing interest in renewable energy sources such as solar, wind, and geothermal energy, as well as the utilization of biomass. Solar and wind energy are expected to gradually replace fossil fuels for heating and electricity generation. However, transitioning away from gasoline, diesel, and jet fuel in the transportation sector poses a greater challenge. While green electricity can be convenient for short-distance transport, it is not feasible for air travel or heavy long-distance transportation. Consequently, a portion of the transportation fuel sector will likely need renewable carbon-based fuels. Biomass-based renewable transportation fuels offer an attractive solution for this purpose.

At present, first-generation biofuels, such as bioethanol derived from sugar/starch crops and biodiesel obtained from oil-bearing seeds, have been successfully commercialized and are readily available in the market. However, ethical concerns related to the competition between food and fuel have hindered the widespread adoption of these initial biofuels. Consequently, there is growing interest in utilizing lignocellulosic biomass (also known as woody biomass) as a non-food carbon resource. Examples of lignocellulosic biomass include wood and agricultural residues like straw, grass, nuts, and corn stover. The precise composition of lignocellulosic biomass varies depending on the source, but it consistently contains three main components. First, cellulose (approximately 40% by weight): A high-molecular-weight polymer composed of glucose monomers. Its chemical formula can be represented as $(C_6H_{10}O_5)_n$, where n ranges from 500 to 4000. Cellulose forms bundled fibers that provide strength to plants and trees. Second, hemicellulose (around 25% by weight) comprising both C_6 and C_5 sugars, hemicellulose has a significantly lower average molecular weight than cellulose. It acts as a binding agent, gluing together cellulose bundles. Third, lignin (approximately 20% by weight) forms a three-dimensional network of highly branched and substituted aromatic molecules. It is covalently bound to the hemicellulose fraction .

Airlines have committed to achieving carbon-neutral growth in international commercial aviation starting in 2021. U.S. airlines have set an ambitious goal to reduce carbon dioxide (CO_2) emissions by 50% by 2050, compared the levels recorded in 2005. Although U.S. airlines have already improved efficiency by 130% since 1978, further enhancements in aircraft and engine efficiency alone may not suffice. To meet the 2050 target, they will need fuels with a lower carbon footprint, commonly known as sustainable aviation fuel (SAF). One of the challenges in providing sustainable aviation fuel (SAF) lies in the expansive and growing jet fuel market. Another significant challenge lies in the current cost of sustainable aviation fuel (SAF), which exceeds that of petroleum-based Jet A fuel. Given that fuel constituents 20%-30% of an airline's operating expenses, this price disparity poses a hurdle. However, research and development (R&D) efforts can help reduce SAF costs. Unlike light-duty vehicles, where electrification is a viable option, the low energy density of even the best batteries severely limits opportunities for electrification in aviation. While progress is being made for smaller aircraft, airlines currently have no alternative but to rely on SAF to operate within a GHG-emission-constrained future.

Jet fuels consist of various components, including n-alkanes, iso-alkanes, cycloalkanes, and aromatics. Aromatics do not burn as cleanly as alkanes, result in higher particulate emissions and have lower specific energy. The n-alkanes are acceptable, they don't meet fluidity and handling properties well. The iso-alkanes possess high specific energy, good thermal stability, and low freezing points. Cycloalkanes brings complementary value to iso-alkanes by providing similar benefits to aromatics, enable fuels to meet density requirements and may potentially offer seal-swelling capacity similar to that provided by aromatics. Combined, iso-alkanes and cycloalkanes can enhance fuel properties by enabling high specific energy and energy density, minimize emission characteristics. This fuel properties could lead to increased range, higher payload capacity, or fuel savings.

Various technologies are available for converting lignocellulosic biomass into transportation fuel. In general, the technologies for converting lignocellulosic biomass into transportation fuel can be categorized into thermochemical processes (gasification, pyrolysis, and liquefaction) and low-temperature processes (hydrolysis, fermentation, and anaerobic digestion). Biomass gasification aims to produce syngas, primarily composed of CO and H_2. Syngas serves as a valuable precursor for liquid fuels, such as those produced via the Fischer-Tropsch process. This work focuses on transforming lignocellulosic biomass into liquid transportation fuels through a process involving fast pyrolysis, followed by catalytic hydrotreatment and refining. During fast pyrolysis, lignocellulosic biomass is rapidly heated in the absence of oxygen to high temperatures (typically around 500°C). Biomass breaks down into vapors, aerosols, and some char and non-condensable gases. To prevent repolymerization of vapor components, the residence time of the vapors is kept short (around 2 seconds or less). The resulting vapors are then rapidly cooled to maximize the yield of a dark brown, viscous liquid known as "fast pyrolysis oil" (or bio-oil). This bio-oil can serve as a green combustion fuel or as an intermediate product for further conversions (such as hydrotreatment). It essentially represents liquefied lignocellulosic biomass. Advantages include easier transport compared to the original solid biomass and a lower ash content.

Catalytic hydrotreatment is a process in which a liquid feed reacts with hydrogen and a catalyst at elevated temperature and pressure. When applied to fast pyrolysis oil, the primary goal is to remove reactive oxygenates from the oil, enhancing its stability and reducing acidity. This hydrotreatment typically occurs in batch, slurry, or packed bed reactors. Achieving significant reductions in oxygen content requires relatively long reaction times in batch reactors (typically around 4 hours) or low space velocities in continuous reactors. Initially, catalytic hydrotreatment of fast pyrolysis oil was likened to hydrodesulfurization (HDS), a well-established process used to reduce Sulphur content in oil refinery streams. The main objective of hydrotreatment is to remove heteroatoms (such as Oxygen (O) in the case of fast pyrolysis oil or Sulphur (S) in fossil feeds) to form water and H_2S, respectively. The fast pyrolysis of biomass is already being commercialized, while the upgrading via bio-oil hydrotreating to transportation fuels has only been demonstrated in the laboratory and on a small engineering development scale. Based on the experiment results, we can explain through the simulation of the process design of pyrolysis oil upgrading via hydrotreating section is revised to incorporate the most recent improvements: a low-temperature stabilizer reactor has been added ahead of the one-stage hydrotreaters to obtain the sustainable aviation fuel.

28a Kinetics and Mechanism of Multiscale Formation and Inhibition in Oilfield Industries

Glavic Tikeri, Ali Alshami

Effective mineral scale control is pivotal across diverse industries that utilize brine, especially oil and gas, to uphold productivity, facility integrity, and safety standards. Among various mitigation strategies to control problematic scale formation, scale inhibitors are a cost-efficient solution. Optimizing their application demands a precise evaluation of its performance, which is this study's goal. We introduce a groundbreaking approach, integrating a novel continuous stirred tank reactor with laser techniques to monitor multiscale formation kinetics in industrial processes and assess inhibitor efficacy. Common scale types such as calcium carbonate and calcium sulfate pose significant challenges in numerous industries grappling with scale formation issues; they impede heat transfer, block major oilfield equipment, and their interaction as mixed scale could cost production facilities millions of dollars. We capture the induction time variations under different replicated oilfield conditions by continuously monitoring laser intensity changes in oversaturated solutions of these salts stirred in the reactor. Inhibitor efficiency is gauged based on how much it delays the induction time. Further insights into inhibition kinetics are gleaned through scanning electron microscopy analysis and X-ray Diffraction, revealing diverse mechanisms and polymorphic scale structures contingent upon different inhibitor compositions. Our findings enhance our understanding of multiscale formation dynamics and inform targeted inhibitor development strategies, promising more effective scale control measures and streamlined industrial operations.

USE OF NANOPARTICLES TO CONTROL FORMATION DAMAGE DUE TO FINES MIGRATION IN THE PRODUCTION - STATE OF THE ART

Edwars A. Naranjo[1,2,3], Lourdes M. Orejuela-Escobar[1,2,4,5], Inés C. Hernández[1,2,5]

[1]Grupo de Ingeniería Circular Aplicada, Departamento de Ingeniería Química, Facultad de Ciencias e Ingeniería, Universidad San Francisco de Quito USFQ. Quito, Ecuador.
[2]Departamento de Ingeniería Química, Facultad de Ciencias e Ingeniería, Universidad San Francisco de Quito USFQ. Quito, Ecuador.
[3]Universidad de Carabobo, UC. Valencia, Venezuela.
[4]Instituto de Energía y Materiales, Universidad San Francisco de Quito, USFQ. Quito, Ecuador.
[5]Instituto de Investigaciones Biológicas y Ambientales BIÓSFERA, Universidad San Francisco de Quito, USFQ. Quito, Ecuador.

Abstract

Fines migration, a common problem in oil production, occurs when fine particles break away from the rock formation and are carried away by the fluid, causing pore clogging and decreased permeability. This phenomenon reduces the wells' productivity and affects the projects' profitability. The present article reviews the use of nanoparticles as a promising solution to control fines migration. The migration mechanisms, the factors that influence it, and the various applications of nanoparticles to inhibit and stabilize fines are explored. Scientific evidence supports the effectiveness of nanoparticles in improving oil production by mitigating damage caused by fines migration.

Introduction

The migration of fine particles occurs when the existing equilibrium in the interaction between rock and reservoir fluid is altered due to viscous and interfacial forces that promote the movement of particles loosely adhered to the pores. This can occur during different phases of crude oil recovery from the reservoir, such as drilling, production, hydraulic fracturing, and workover operations[1-4]. It usually occurs during the production stage, when the fluid within the reservoir moves toward the wellbore and the fluid velocity increases. As the fluid reaches the critical flow rate, the fine particles (density <1400 kg/m^3, diameter <37 microns)[5] present in the formation break off and are carried away by the fluid. However, minerals in the porous medium (clays, amorphous silica, quartz, etc.) can also migrate and cause damage to the formation.

Fine particles deposited along flow channels can obstruct fluid flow within the reservoir. These particles are not always small enough to pass through the pore throat, becoming trapped and accumulating in smaller pore spaces or thinner pore throats. This causes clogging of the pores and decreased permeability, thus reducing the production rate. This phenomenon of clogging or migration of fines is one of the main reasons for the decrease in oil and gas wells' productivity, affecting the projects' profitability[1-4].

A *porous medium* is a solid material containing an interconnected network of void spaces or pores. These pores can vary in size, shape, and connectivity, creating an intricate labyrinth through which fluids can flow. The complexity of a porous medium lies in the diversity of forces that govern the movement of particles within it. These forces include hydrodynamic interactions between the particles and the fluid, surface forces between the particles and the porous matrix, and electrostatic effects that attract or repel the particles. Furthermore, the irregular geometry of pores and throats can create trapping zones and alter flow patterns, making it even more challenging to predict the behavior of particles in the porous medium[6-7].

The migration of fines involves several main steps (Figure 1a): the presence of fine particles in the rock or weakly cemented to the porous matrix that can be released and mobilized[8], the detachment mechanism that leads to the destruction of the fines; and the trapping of fines at specific locations in the formation due to wettability, ionic attraction, or bridging[9].

This article aims to provide a bibliographic review of the use of nanoparticles to control formation damage derived from the migration of fines, especially in the production stage. It considers the forms and techniques applied for its control, compiles the existing scientific evidence on the topic, and discusses research recommendations.

The study of particle movement in a porous medium is complex due to the various forces interacting in this environment. A porous medium consists of a solid part with many microscopic pores inside and throats or spaces between them. This network of narrow conduits is the path through which the fluid can cross the medium[10].

Oil fields are initially produced by natural energy, which decreases with increasing production and finally reduces at the end of primary production. In other words, the extraction of oil fluids and connate formation water leads to an increase in local formation pressure or effective stress, which acts on the reservoir rock, causing a decrease in local pore pressure to counteract these forces. According to the theory of fines migration[11], there are main forces that act on each fine particle moving along the pore wall (Figure 1b and Figure 1c), and other forces responsible for separating the fines are the forces Van der Waals attraction, Born repulsion forces, the double layer phenomenon, and hydrodynamic forces that overcome the detachment of fine particles in the porous medium. When the total interaction energy between the particle and the porous surface becomes positive, the repulsive forces are more significant than the attractive forces, which occur when the fines are released[9].

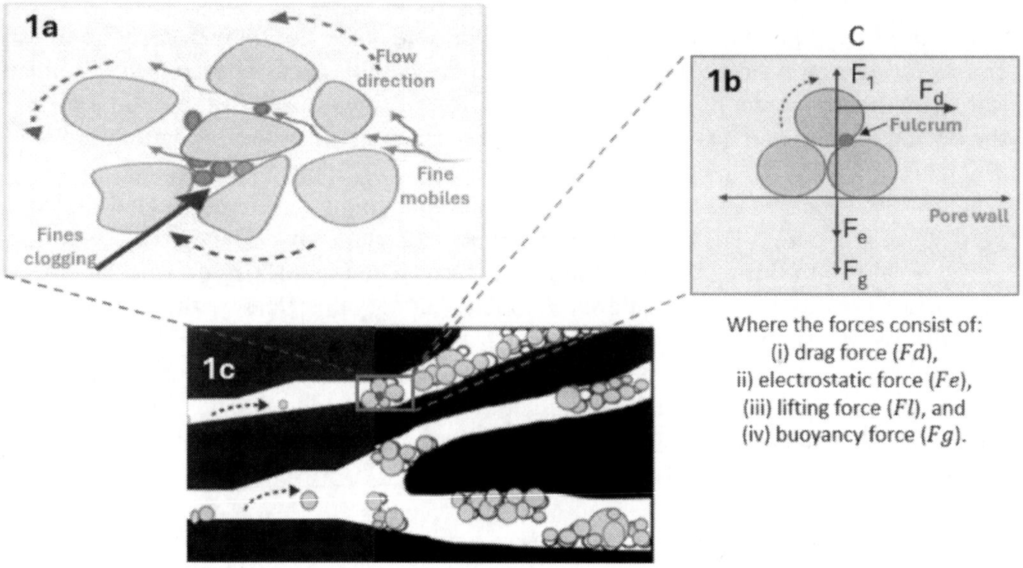

Figure 1. Schematics of pore restriction and particle force balance on the surface of pore channel within porous media[47]. 1a) Pore restriction due to migration of fines. 1b) Pore restriction due to migration of fines. 1c) Particle aggregates as internal cake station on the pore wall surface

During production, various mechanisms occur, among which the adsorption of particles on the outside of the rock or its surface stands out due to Brownian movement and the electrostatic interaction between the migrating particles and the surface of the pores[2]. Several researchers have studied the effect of different factors on fines production, such as salinity, flow rate, pH, residual oil saturation, temperature, wettability, crude oil polarity, and fractional flow of water and crude oil[12].

Fines migration mechanisms

Fines migration in oil fields is a complex phenomenon involving interrelated processes. The three main steps that characterize this phenomenon are described below:

Presence of fines

The first crucial step in fine migration is the existence of these particles within the rock formation. Fines can originate from various sources, including:

Clay minerals: constitute one of the primary sources of fines in deposits. Minerals such as kaolinite, illite, and montmorillonite are particularly susceptible to migration due to their small size and swelling properties[13-15].

Other minerals: in addition to clays, other minerals such as silica, quartz, and feldspars can also contribute to the fine population in the formation[16-17].

Corrosion and degradation products: the interaction of reservoir fluids with rock and wellbore construction materials can generate fine particles as corrosion or degradation products[18].

These fines can be present in two primary forms: [19-22]

Adhered to the porous matrix: fines may be weakly cemented or adhered to the porous structure of the rock, making them susceptible to dislodgement under certain conditions.

Free in the pores: some fines can exist as free particles within the pore spaces of the formation, which facilitates their mobilization by fluid flow.

Detachment mechanism

Once the fines are present in the formation, the next step is detachment from the porous matrix or their mobilization from the pores. Several mechanisms can contribute to this process [19] [23-25]:

Hydrodynamic forces: the flow of fluids through the formation exerts drag forces on the fines, which can overcome the adhesion forces and cause their detachment.

Changes in fluid chemistry: variations in the reservoir fluids' salinity, pH, or ionic composition can alter the interaction forces between the fines and the porous matrix, facilitating their release.

Changes in effective stress: decrease pore pressure during production can increase the rock's stress, weakening the pore structure and favoring the release of fines.

Physicochemical interactions: phenomena such as the expansion and contraction of clays, the adsorption and desorption of ions, and the alteration of wettability can influence the stability of the fines and promote their release.

Fines trapping

Once the fines have been dislodged, they are transported by the fluid flow until they find a trapping point within the formation. The main trapping mechanisms include[15] [21] [26]:

Mechanical blocking: fines can become physically trapped in narrower pore throats or smaller pore spaces, reducing formation permeability.

Adhesion: attractive forces between fines and the rock surface, such as Van der Waals forces and electrostatic interactions, can cause their adhesion and accumulation in the pores.

Bridging: fines can form bridges or structures between rock grains, further clogging flow channels and reducing permeability.

Factors that influence the migration of fines

Various interconnected factors influence the migration of fines in oil fields. These factors can affect the stability of the fines, their release from the porous matrix, and their subsequent transport and entrapment. Some of the most critical factors influencing fines migration are highlighted below, emphasizing salinity, flow rate, pH, and other relevant factors:

Salinity

Reservoir fluid salinity plays a crucial role in fines migration, particularly in formations containing salinity-sensitive clays such as montmorillonite[24] [27] :

Low salinity: clays can swell and weaken the porous structure in low salinity conditions, facilitating fine release. Furthermore, low salinity can reduce the electrostatic attraction forces between the fines and the porous matrix, promoting their mobilization.

High salinity: in contrast, high salinity can compress clays and strengthen the porous structure, making it challenging to release fines. Furthermore, high salinity can increase electrostatic attraction forces, favoring the adhesion of fines to the porous matrix.

Flow

The flow rate or flow rate of fluids in the formation significantly influences the migration of fines[28]:

Low flow: at low flow rates, the hydrodynamic forces are weak and may not be sufficient to overcome the adhesion forces of the fines, limiting their release and transport.

High flow rate: as the flow rate increases, the hydrodynamic forces intensify, which can cause the release of fines, even those firmly adhered to the porous matrix. Additionally, high flow can transport fines greater distances within the formation, increasing the potential for damage.

pH

The pH of reservoir fluids can also affect the migration of fines, especially in formations containing pH-sensitive minerals[27] [29]:

Low pH: an acidic pH can dissolve some minerals and weaken the pore structure, facilitating fines' release.

High pH: an alkaline pH can promote mineral precipitation and scale formation, blocking pores and trapping mobilized fines.

Other factors

In addition to salinity, flow rate, and pH, other factors can also influence the migration of fines[27] [30-34]:

Temperature: can affect fluid and rock properties and physicochemical interactions between fines and the porous matrix.

Crude oil composition: the crude oil composition, especially its asphaltene and resin content, can influence rock wettability and fines adhesion.

Residual oil saturation: residual oil in the formation can alter capillary forces and wettability, affecting fines mobilization and trapping.

Production operations: drilling, hydraulic fracturing, and waterflooding can induce changes in reservoir conditions and trigger fines migration.

Applications of nanoparticles in the control of fines migration

Types and mechanisms of action

Nanoparticles (NPs) have emerged as a promising tool for addressing fine migration in oil fields. Their nanometer size and unique properties allow them to interact effectively with fines and the porous matrix, offering diverse mechanisms of action to control migration and mitigate formation damage. The main types of nanoparticles used, and their mechanisms of action are detailed below:

Silica nanoparticles (SiO_2): are widely used due to their abundance, low cost, and biocompatibility. Their primary mechanism of action is adhesion and strengthening of the porous matrix. Silica NPs adhere to the fines' surface and the porous matrix, forming a protective layer that prevents the detachment of the fines. In addition, they can penetrate microfractures and pores, strengthening the structure of the rock and reducing its susceptibility to damage[21] [28] [35-36].

Alumina nanoparticles (Al_2O_3): also effective in controlling fines migration, play a crucial role in altering the wettability of the rock. Their primary action mechanism is modifying the rock's surface from hydrophilic (affinity for water) to more hydrophobic (affinity for oil). This alteration reduces the interaction between aqueous fluids and fines, decreasing their detachment and transport [28] [37-38].

Polymeric nanoparticles: offer a wide range of possibilities due to the diversity of polymers available. Their mechanism of action can vary depending on the type of polymer used. Some polymers can act as fines migration inhibitors, adsorbing on the surface of the fines and preventing their detachment. Other polymers can form a three-dimensional network within the pores, trapping the mobilized fines and preventing their further migration[39-40].

Metal nanoparticles: although less common, metallic nanoparticles, such as those of iron or nickel, have also been investigated to control fines migration. Their mechanism of action may include altering the magnetic properties of the fines or catalyzing chemical reactions that promote the stabilization of the fines.

Other types of nanoparticles

In addition to those mentioned above, other types of nanoparticles have been explored, such as carbon, titanium oxide, and zeolite, each with their specific mechanisms of action[41]. The literature reports various methods to address formation damage problems resulting from fines migration. These methods have been applied; However, some have failed, others have been ineffective, and others include nanoparticles (NPs).

The latter has reported good results as inhibitors and stabilizers of the migration of fines due to their unique characteristics, such as their high surface/volume ratio, which gives them the particularity of having a large contact area and, consequently, an increase in the surface energy of the particle[42].

Furthermore, nanoparticles are highly resistant to degradation in media with high salinities and temperatures[43]. This suggests that nanoparticles can adapt to the conditions of the reservoir, allowing an adequate flow of them through the reservoir's pores and facilitating treatment within the porous medium to control damage to the formation due to particle migration[1]. Table 1, summarizes some research on nanoparticles that inhibit and stabilize formation damage due to fines migration [1] [28] [35] [44-47].

Table 1. Summarized some research on Nanoparticles that Inhibit and stabilize formation damage due to fines migration

Author(s) and year	Type of nanoparticles	Action mechanism	Evaluation conditions	Key results
Cárdenas et al. (2024) [44]	Nanofluids	Alteration of wettability	Heavy oil, field test	Improvement in wettability and increase in oil production.
Agi et al. (2024) [28]	Mesoporous silica	Fines trapping	Box-Behnken experimental design	Optimization of the fines trapping process.
Gallego (2024) [45]	Nano-chemicals dispersed in gas	Increased recovery factor	Injection of nano chemicals in the reservoir	Improvement in the oil recovery factor
Muneer et al. (2024)[35]	Nanofluids	DLVO modeling to predict critical salt concentration	Evaluation of fines migration before and after treatment	Prediction of the critical salt concentration to initiate the migration of fines
Zhao et al. (2021)[1]	five types of nanoparticles, including Al_2O_3 (aluminum oxide)	Control of fines migration	Unconsolidated sandstone formations	Study of the mechanism and effect of nanoparticles in the control of fines migration
Mansour et al. (2021)[46]	Nanotechnology	Prevention of migration of fines	Application during production	Use of nanotechnology to prevent the migration of fines
Tangparitkul et al. (2020)[47]	not specified	Analytical study of fines migration	Reservoir depletion and injection of low salinity water	Analysis of the migration of fines and the decrease in permeability

According to the laboratory investigations and field applications in Table 1, no evidence that using NPs causes permeability deterioration. NPs represent an advantage compared to other chemicals and techniques used to control the migration of fines in formations. Points expanded below:

Scientific evidence

Efficiency of nanoparticles in controlling the migration of fines

Numerous laboratory studies and field applications have supported nanoparticles' effectiveness in controlling fine migration, demonstrating their ability to reduce formation damage and improve healthy productivity. Below are some notable examples of scientific evidence[3] [41] [42] [48-52]:

Laboratory investigations

Silica nanoparticles: studies have shown that treatment with silica nanoparticles can significantly reduce water permeability in sandstone cores, indicating a decrease in fines migration. Furthermore, it has been observed that silica nanoparticles can improve the mechanical strength of the rock, reducing its susceptibility to stress damage:

Alumina nanoparticles: laboratory experiments have shown that alumina nanoparticles can effectively alter rock wettability, reducing the interaction between water and fines. This results in a decrease in the release and transport of fines, which protects formation permeability.

Polymeric nanoparticles: studies have shown that polymeric nanoparticles can act as effective inhibitors of fines migration, adsorbing on the surface of the fines and preventing their detachment. Additionally, some polymers can form a three-dimensional network within the pores, trapping mobilized fines and preventing their further migration.

Field applications

Silica nanoparticles: silica nanoparticles injection into production wells effectively reduces water production and increases oil production. This suggests nanoparticles can mitigate formation damage caused by fines migration and improve healthy productivity.

Alumina nanoparticles: the application of alumina nanoparticles in reservoirs with water production problems has shown promising results in reducing water and increasing oil production. This indicates nanoparticle-induced wettability alteration can effectively control fine migration under field conditions.

Polymeric nanoparticles: the injection of polymeric nanoparticles in wells with fines migration problems has proven to improve productivity and reduce formation damage. The ability of polymers to inhibit fine migration and form trapping networks contributes to these positive results.

Advantages of nanoparticles in controlling the migration of fines

Nanoparticles present several distinctive advantages that make them an attractive tool for addressing the problem of fines migration in oil fields. These advantages primarily derive from their unique physicochemical properties and ability to effectively interact with fines and the porous matrix. Some of the most essential advantages of nanoparticles in this context are highlighted below:

High surface/volume ratio: one of the most notable characteristics of nanoparticles is their high surface/volume ratio. This means that, for the same mass of material, nanoparticles have a much larger surface area than larger particles. This large surface area gives them a greater capacity for interaction with the fines and the porous matrix, which translates into greater efficiency in the adhesion mechanisms, alteration of wettability, and formation of trapping networks [53-54].

Resistance to degradation: nanoparticles exhibit remarkable resistance to degradation in hostile environments, such as oil fields. This resistance is partly due to its compact crystalline structure and the absence of surface defects that can act as starting points for degradation. Furthermore, some nanoparticles can be coated with protective materials that further increase their stability under conditions of high temperature, high salinity, and aggressive chemicals [21] [53].

Adaptability to reservoir conditions: nanoparticles can be designed and modified to adapt to the specific conditions of each reservoir. This includes selecting materials and coatings compatible with reservoir fluids and rock and optimizing the size and shape of nanoparticles to facilitate their transport and penetration into the formation. This adaptability allows a more effective and personalized application of nanoparticles to control the migration of fines in different reservoirs[21] [55].

Other advantages

In addition to the advantages mentioned above, nanoparticles also offer other potential benefits, such as [39] [48] [56-57]:

Low toxicity: many nanoparticles, especially silica and polymeric ones, are biocompatible and have low toxicity, which reduces the environmental impact of their application in oil fields.

Ease of handling: nanoparticles are typically easy to handle and disperse in treatment fluids, simplifying their application in field operations.

Cost-effectiveness: due to their high efficiency and durability, nanoparticles can offer a more cost-effective solution in the long term compared to other fines migration control technologies.

Discussion

The promising future of nanoparticles in oil production:

Fines migration in oil fields is a dynamic and complex process involving various factors, such as salinity, flow, pH, and oil properties. Understanding the underlying mechanisms of fines presence, detachment, and entrapment is essential to develop effective formation damage control and mitigation strategies.

Nanoparticles have emerged as a promising tool to address this challenge, offering a variety of mechanisms of action, such as inhibiting fines detachment, stabilizing the porous matrix, and altering rock wettability. Scientific evidence, both from laboratory research and field applications, strongly supports the effectiveness of nanoparticles in reducing water production, increasing oil production, and improving overall well productivity.

The inherent advantages of nanoparticles, such as their high surface-to-volume ratio, resistance to degradation, and adaptability to reservoir conditions, make them an attractive and versatile solution. Furthermore, their low toxicity and ease of handling make them even more promising from an environmental and operational perspective.

While significant progress has been made, continued research and development in this field are essential to unlocking the full potential of nanoparticles in oil production. Optimizing the selection and application of nanoparticles for different types of formations and reservoir conditions and exploring new materials and mechanisms of action are critical areas for future research. Developing "smart" nanoparticles that respond to reservoir-specific stimuli and creating more sophisticated predictive models also represent promising opportunities to improve the efficiency and personalization of treatments.

Ultimately, nanoparticles' success in controlling fine migration will depend on a deep understanding of the mechanisms involved and the ability to tailor solutions to the specific conditions of each reservoir. Interdisciplinary collaboration and knowledge transfer between academia and industry will be essential to drive innovation and ensure the successful application of this technology in the field.

Conclusions

Fine particles in the porous medium are susceptible to migration and can become trapped within the matrix, reducing porosity and permeability and causing formation damage. As reservoir pressure decreases during production, the effective stress on the rock increases, which can induce greater particle detachment.

The literature review demonstrates that nanoparticles offer a promising solution to control the fines migration and improve the oil production stage. However, it is essential to recognize that the effectiveness of this method may vary depending on the specific wellbore and formation conditions and the implementation strategy.

Despite the advances, the future of nanoparticles in oil production is even more promising. Continuous research in optimizing their design, application, and monitoring will allow the development of tailored solutions for each reservoir, maximizing hydrocarbon recovery and extending the useful life of the wells. This approach represents a technological advance and a step towards a more efficient and sustainable oil industry.

Research recommendations

Expanding the horizon of nanoparticles in the fight against fines migration. Although nanoparticles have shown great potential in controlling fines migration, there are still numerous areas of research that can boost their application and effectiveness in the oil industry. Some key recommendations for future research are suggested below[21] [28] [42] [58-60].

Optimization of the selection and application of nanoparticles

Reservoir characterization: delve into the characterization of different formations and reservoir conditions to identify the key factors that influence the migration of fines and select the most appropriate nanoparticles for each case. A thorough understanding of the formation and reservoir fluid properties is essential to selecting the most suitable nanoparticles and designing effective treatment strategies. This will maximize the nanoparticles' efficiency and minimize the risk of adverse effects.

Custom nanoparticle design: develop nanoparticles with specific physicochemical properties (size, shape, surface charge, functionalization) to optimize their interaction with fines and the porous matrix in different reservoir conditions. Customizing nanoparticles to suit specific reservoir conditions can significantly improve their performance and reduce costs by avoiding ineffective or incompatible treatments.

Injection strategies: investigate and optimize nanoparticle injection strategies, including concentration, flow rate, contact time, and compatibility with other treatment fluids, to maximize their distribution and formation effectiveness. Optimization of injection strategies is crucial to guarantee a uniform and effective distribution of nanoparticles in the formation, maximizing their interaction with the fines and the porous matrix.

Monitoring and evaluation: develop real-time monitoring and evaluation techniques to track the distribution and performance of nanoparticles in the reservoir, allowing adjustments to the treatment strategy if necessary. Real-time monitoring of nanoparticle performance allows treatment effectiveness to be evaluated and adjustments made if necessary, optimizing control of fines migration and improving well productivity.

New materials and mechanisms of action

Hybrid nanoparticles: explore the development of hybrid nanoparticles that combine different materials and functionalities to achieve more effective and multifaceted control of fines migration. Combining different materials and functionalities in hybrid nanoparticles can offer complete and synergistic control of fines migration, addressing multiple formation damage mechanisms.

Smart nanoparticles: investigate the design of "smart" nanoparticles that respond to reservoir-specific stimuli, such as changes in temperature, salinity, or pH, releasing active agents or modifying their properties to adapt to changing conditions. The development of nanoparticles that respond to reservoir-specific stimuli would allow a controlled release of active agents or an adaptation of their properties depending on changing conditions, improving the efficiency and durability of the treatment.

Innovative mechanisms of action: explore new mechanisms of action for controlling fines migration, such as modifying properties, rheological conditions of the reservoir fluids, or the induction of chemical reactions that promote the stabilization of the fines. Exploring new mechanisms of action can lead to developing more effective and versatile solutions to control fines migration, especially in complex or challenging reservoirs.

Modeling and simulation

Predictive models: develop more sophisticated mathematical and computational models to predict fines migration and evaluate the effectiveness of different nanoparticle treatments under various reservoir conditions. Developing more accurate predictive models will allow simulation of the behavior of nanoparticles in the reservoir, optimize treatment strategies, and reduce costs by avoiding unnecessary field tests.

Field-scale simulations: perform field-scale simulations to evaluate the long-term impact of nanoparticle treatments on well productivity and hydrocarbon recovery.

Economic and environmental considerations

Cost-benefit analysis: perform detailed cost-benefit analyses to evaluate nanoparticle treatments' economic viability compared to other fines migration control technologies. Assessing the economic viability of nanoparticle treatments is crucial to ensure their large-scale adoption and application in the petroleum industry.

Environmental impact: investigate the potential environmental impact of nanoparticles in the oil industry and develop strategies to minimize any adverse effects on ecosystems. Research into the environmental impact of nanoparticles is essential to ensure the sustainability of their use in the oil industry and minimize any adverse effects on ecosystems.

Collaboration and knowledge transfer

Interdisciplinary collaboration: foster collaboration between researchers from different disciplines, such as materials science, petroleum engineering, chemistry, and geology, to drive innovation and develop new nanoparticle-based solutions. Collaboration between different disciplines is essential to drive innovation and the development of new solutions based on nanoparticles, taking advantage of knowledge and experience from various fields.

Knowledge transfer: facilitate the transfer of knowledge and technologies between academia and industry to accelerate the adoption and application of nanoparticle-based solutions in the field. The effective transfer of knowledge and technologies between academia and industry accelerates the adoption and application of innovative solutions, such as those based on nanoparticles, in the field, benefiting both parties and society.

References

1. Zhao, X., Qiu, Z., Gao, J., Ren, X., Li, J., & Huang, W. (2021). Mechanism and effect of nanoparticles on controlling fines migration in unconsolidated sandstone formations. *SPE Journal*, *26*(06), 3819-3831.
2. Awan, F. U. R., Arif, M., Iglauer, S., & Keshavarz, A. (2022). Coal fines migration: A holistic review of influencing factors. *Advances in Colloid and Interface Science*, *301*, 102595.
3. Fuentes, J. F., Montes, D., Lucas, E. F., Montes-Paez, E. G., Szklo, A., & Guerrero-Martin, C. A. (2022). Nanotechnology applied to the inhibition and

remediation of formation damage by fines migration and deposition: A comprehensive review. *Journal of Petroleum Science and Engineering, 216*, 110767.

4. Ogolo, N. A., & Onyekonwu, M. O. (2021). Particle Migration in Hydrocarbon Reservoirs During Production: Problems and Trigger Factors. *Petroleum & Coal, 63*(1).
5. Civan, F. (2023). *Reservoir formation damage: fundamentals, modeling, assessment, and mitigation*. Gulf Professional Publishing.
6. Sivakumar, N., Noeiaghdam, S., & Fernandez-Gamiz, U. (2024). Multiple shape factor effects of nanofluids on marangoni mixed convection flow through porous medium. *Results in Engineering, 23*, 102512.
7. Fan, J., Chen, Z., Fu, S., Zhu, Y., & Wang, L. (2024). Numerical study of gas–solid transport characteristics within vortex region induced by porous media using lattice-Boltzmann based discrete particle simulation. *Powder Technology, 434*, 119336.
8. Zhang, F., Wang, T., Liu, F., Peng, M., Bate, B., & Wang, P. (2022). Hydro-mechanical coupled analysis of near-wellbore fines migration from unconsolidated reservoirs. *Acta Geotechnica, 17*(8), 3535-3551.
9. Yang, Y., Yuan, W., Hou, J., & You, Z. (2022). Review on physical and chemical factors affecting fines migration in porous media. *Water Research, 214*, 118172.
10. De Castro, A. R., & Goyeau, B. (2021). A pore network modelling approach to investigate the interplay between local and Darcy viscosities during the flow of shear-thinning fluids in porous media. *Journal of colloid and interface science, 590*, 446-457.
11. Ruiz Rozo, D. D. P. (2020). Evaluación del efecto de las nanopartículas de alúmina y sílice presentes en el filtrado del lodo para la inhibición de finos y la preservación de la humectabilidad al agua en un campo de los Llanos Orientales.
12. Ali, A. S., Salem, A., & Attia, A. M. (2022). Mitigation of fines migration by using nano-fluid to alleviate formation damage in Abu Rawash G. *Petroleum & Coal, 64*(2).
13. Lei, J., Wang, Y., & Guo, W. (2024). Fines migration characteristics of illite in clayey silt sediments induced by pore water under the depressurization of natural gas hydrate. *Fuel, 365*, 131236.
14. Lei, J., Wang, Y., & Guo, W. (2024). Migration characteristics and geohazards risk analysis of swelling clay in clayey silt hydrate sediments. *Gas Science and Engineering, 126*, 205347.
15. Wang, Z., Yin, H., Tang, H., Hou, Y., Yu, H., Liu, Q., & Tang, H. (2024). Analysis of damage mechanisms and controlling factors of fine particle migration in unconsolidated sandstone reservoirs based on reservoir classification. *Energy Science & Engineering, 12*(7), 2879-2898.
16. Lyu, S., Lu, A., Li, R., Chen, X., Chen, Z., Chen, L., ... & Xiong, Z. (2024). Production of Coal Fines in Deep Coal Seam via Morphologic, Flow-Channel, Mineral Analyses: Evidence from Coalbed Methane Wells in the Qinshui Basin. *ACS omega*.
17. Li, J., Yang, Z., Zhang, W., Hu, W., Wang, Y., Wu, J., ... & Lin, Z. (2024). Inertial migration of fine mineral particles in a curved microfluidic channel: Demystifying the role of non-neutrally buoyant particles. *Separation and Purification Technology, 334*, 126026.

18. Lu, D., Jiang, X., Qu, F., & Huo, Y. (2024). Mitigating sulfate ions migration in concrete: A targeted approach to address recycled concrete aggregate's impact. *Journal of Cleaner Production*, *442*, 141135.
19. Liu, P., Nie, S., Wang, W., Zhang, S., Bate, B., & Chen, Y. (2024). CFD-DEM study on transport and retention behaviors of nZVI-clay colloids in porous media. *Journal of Hazardous Materials*, *465*, 133048.
20. Zeng, L., Yuan, C., Xiang, T., Guan, X., Dai, L., Xu, D., ... & Tian, C. (2024). Research on the Migration and Adsorption Mechanism Applied to Microplastics in Porous Media: A Review. *Nanomaterials*, *14*(12), 1060.
21. Saghandali, F., Salehi, M. B., Pahlevani, H., Taghikhani, V., Riahi, S., Ebrahimi, M., ... & Roomi, A. (2024). Fabrication of a hydrogel reinforced with titanium nanoparticles to reduce fine migration and remediation of formation damage during low salinity waterflooding. *Geoenergy Science and Engineering*, 213173.
22. Kou, X., Zhang, H., Li, X. S., Chen, Z. Y., & Wang, Y. (2024). Methane hydrate phase transition in marine clayey sediments: Enhanced structure change and solid migration. *Applied Energy*, *368*, 123485.
23. Nguyen, C., Loi, G., Russell, T., Yang, Y., Zulkifli, N. N., Amir, M. I., ... & Zeinijahromi, A. (2024). Physics mechanisms of fines detachment and migration during CO2-water corefloods. *arXiv preprint arXiv:2403.14343*.
24. Dadjou, H., Miri, R., Salimi, M., & Nouri, A. (2024). Near-Wellbore Salinity Effect on Sand Control Plugging by Fines Migration in Steam-Assisted Gravity Drainage Producer Wells. *SPE Journal*, *29*(06), 2808-2821.
25. Kozhevnikov, E. V., Turbakov, M. S., Riabokon, E. P., & Gladkikh, E. A. (2024). Apparent permeability evolution due to colloid migration under cyclic confining pressure: On the example of porous limestone. *Transport in Porous Media*, *151*(2), 263-286.
26. Lei, J., Chen, Z., Zhao, J., Wang, Y., & Guo, W. (2024). Effect of decrease in pore water salinity induced by hydrate decomposition in clayey silt sediment on the destabilization-migration-clogging characteristics of illite. *Energy*, 132887.
27. Sadeghein, A., Abbaslu, A., Riahi, S., & Hajipour, M. (2024). Comprehensive Analysis of Fine Particle Migration and Swelling: Impacts of Salinity, pH, and Temperature. *Geoenergy Science and Engineering*, 213044.
28. Agi, A., Junin, R., Jaafar, M. Z., Amin, N. A. S., Sidek, M. A., Nyakuma, B. B., ... & Mamah, S. (2024). Designed mesoporous silica nanoparticles to mitigate against reservoir fines migration. *Biomass Conversion and Biorefinery*, *14*(2), 2677-2692.
29. Zhao, S., Zhao, Y., Cui, Z., Zhang, H., & Zhang, J. (2024). Effect of pH, Temperature, and Salinity Levels on Heavy Metal Fraction in Lake Sediments. *Toxics*, *12*(7), 494.
30. He, J., Huang, X., & Cao, P. (2024). Fine particle migration in a gas hydrate sand: Single-and two-phase fluid using a device for observation at the pore scale. *Journal of Marine Science and Engineering*, *12*(1), 109.
31. Tao, H. U., Jiang, F., Xiongqi, P. A. N. G., Yuan, L. I. U., Guanyun, W. U., Kuo, Z. H. O. U., ... & Qingyang, M. E. N. G. (2024). Identification and evaluation of shale oil micro-migration and its petroleum geological significance. *Petroleum Exploration and Development*, *51*(1), 127-140.
32. Cao, G., Lin, M., Zhang, L., Ji, L., & Jiang, W. (2024). Numerical simulation of the dynamic migration mechanism and prediction of saturation of tight sandstone oil. *Science China Earth Sciences*, *67*(1), 179-195.

33. Askar, N., Muneer, R., & Pourafshary, P. (2023). Effect of wettability and phase distribution on critical salt concentration for fines migration initiation in sandstone reservoirs. *Journal of Molecular Liquids*, *390*, 123113.
34. Russell, T., Nguyen, C., Loi, G., Mohd Shafian, S. R., Zulkifli, N. N., Zeinijahromi, A., & Bedrikovetsky, P. (2024). Effects of fines migration and reservoir heterogeneity on well productivity: analytical model and field cases. *Journal of Energy Resources Technology*, 1-24.
35. Muneer, R., Pourafshary, P., & Hashmet, M. R. (2024, February). Fines Migration Control in Sandstone Reservoirs: DLVO Modeling for Critical Salt Concentration and Critical Flow Rate Prediction. In *SPE International Conference and Exhibition on Formation Damage Control* (p. D011S001R005). SPE.
36. Alkalbani, A. M., & Chala, G. T. (2024). A comprehensive review of nanotechnology applications in oil and gas well drilling operations. *Energies*, *17*(4), 798.
37. Behera, U. S., Poddar, S., Deshmukh, M. P., Sangwai, J. S., & Byun, H. S. (2024). Comprehensive Review on the Role of Nanoparticles and Nanofluids in Chemical Enhanced Oil Recovery: Interfacial Phenomenon, Compatibility, Scalability, and Economic Viability. *Energy & Fuels*.
38. Anvari, S., & Turzo, Z. (2024). An Experimental Assessment of the Impact of TiO2 and Al2O3 Nanoparticles and Tragacanth on the Augmentation of Oil Recovery in Limestone Reservoirs. *Rudarsko-geološko-naftni zbornik*, *39*(3), 59-76.
39. Yusuf, E. O., Amber, I., Officer, S., & Oluyemi, G. F. (2024). Transport of nanoparticles in porous media and associated environmental impact: a review. *Journal of engineering research*.
40. Giraldo Muñoz, M. A. *Desarrollo de un nanofluido de fractura con doble propósito: aumento en la movilidad de crudos pesados y reducción del daño de formación* (Doctoral dissertation, Universidad Nacional de Colombia).
41. Díez, R., Medina, O. E., Giraldo, L. J., Cortés, F. B., & Franco, C. A. (2020). Development of nanofluids for the inhibition of formation damage caused by fines migration: Effect of the interaction of quaternary amine (CTAB) and MgO nanoparticles. *Nanomaterials*, *10*(5), 928.
42. Ngata, M. R., Yang, B., Aminu, M. D., Iddphonce, R., Omari, A., Shaame, M., ... & Yanyi-Akofur, D. (2021). Review of developments in nanotechnology application for formation damage control. *Energy & Fuels*, *36*(1), 80-97.
43. Panchal, H., Patel, H., Patel, J., & Shah, M. (2021). A systematic review on nanotechnology in enhanced oil recovery. *Petroleum Research*, *6*(3), 204-212.
44. Cárdenas, J. C., Jaimes, M., Dorado, R., Galvis, H., Conde, O. R., & Prada, A. (2024, February). Application of Nanofluids for Altering Wettability in Heavy Oil: A Field Test. In *SPE International Conference and Exhibition on Formation Damage Control* (p. D021S009R002). SPE.
45. Gallego Arias, J. F. (2022, September). Increase in the Oil Recovery Factor through the Injection of Nano-Chemicals Dispersed in Gas. In *SPE Annual Technical Conference and Exhibition?* (p. D011S999R005). SPE.
46. Mansour, M., Eleraki, M., Noah, A., & Moustafa, E. (2021). Uso de la nanotecnología para evitar la migración de finos durante la producción. Petróleo, 7(2), 168-177.
47. Tangparitkul, S., Saul, A., Leelasukseree, C., Yusuf, M., & Kalantariasl, A. (2020). Fines migration and permeability decline during reservoir depletion

coupled with clay swelling due to low-salinity water injection: An analytical study. *Journal of Petroleum Science and Engineering*, *194*, 107448.

48. Franco, C. A., Franco, C. A., Zabala, R. D., Bahamón, I., Forero, A., & Cortés, F. B. (2021). Field Applications of nanotechnology in the oil and gas industry: Recent advances and perspectives. *Energy & Fuels*, *35*(23), 19266-19287.
49. Ibrahim, M. A., Jaafar, M. Z., Yusof, M. A. M., & Idris, A. K. (2022). A review on the effect of nanoparticle in drilling fluid on filtration and formation damage. *Journal of Petroleum Science and Engineering*, *217*, 110922.
50. Ngata, M. R., Yang, B., Aminu, M. D., Emmanuely, B. L., Said, A. A., Kalibwami, D. C., ... & Nyakilla, E. E. (2022). Minireview of formation damage control through nanotechnology utilization at fieldwork conditions. *Energy & Fuels*, *36*(8), 4174-4185.
51. Kalhor Mohammadi, M., Riahi, S., & Boek, E. S. (2023). An insight review on formation damage induced by drilling fluids. *Reviews in Chemical Engineering*, *39*(3), 387-415.
52. Mady, M. F., & Kelland, M. A. (2020). Review of nanotechnology impacts on oilfield scale management. *ACS Applied Nano Materials*, *3*(8), 7343-7364.
53. Hassani, S. S., Daraee, M., & Sobat, Z. (2020). Advanced development in upstream of petroleum industry using nanotechnology. *Chinese Journal of Chemical Engineering*, *28*(6), 1483-1491.
54. Mao, Z., Cheng, L., Liu, D., Li, T., Zhao, J., & Yang, Q. (2022). Nanomaterials and technology applications for hydraulic fracturing of unconventional oil and gas reservoirs: A state-of-the-art review of recent advances and perspectives. *ACS omega*, *7*(34), 29543-29570.
55. Madadizadeh, A., Sadeghein, A., & Riahi, S. (2022). A comparison of different nanoparticles' effect on fine migration by low salinity water injection for oil recovery: Introducing an optimum condition. *Journal of Energy Resources Technology*, *144*(1), 013005.
56. Abdullah, M. M., & Al-Lohedan, H. A. (2024). Nanomaterials in petroleum industry. In *Handbook of Nanomaterials, Volume 1* (pp. 617-648). Elsevier.
57. Candido, J. D. C., Weschenfelder, S. E., & Ferraz, H. C. (2024). A review on the synthesis and application of magnetic nanoadsorbents to the treatment of oilfield produced water. *Brazilian Journal of Chemical Engineering*, *41*(1), 1-21.
58. Giraldo, L. J., Diez, R., Acevedo, S., Cortés, F. B., & Franco, C. A. (2021). The effects of chemical composition of fines and nanoparticles on inhibition of formation damage caused by fines migration: Insights through a simplex-centroid mixture design of experiments. *Journal of Petroleum Science and Engineering*, *203*, 108494.
59. Contreras-Mateus, M., Hethnawi, A., Mheibesh, Y., Montoya, T., Hashlamoun, K., Bakir, M., ... & Nassar, N. N. (2022). Applications of Nanoparticles in Energy and the Environment: Enhanced Oil Upgrading and Recovery and Cleaning up Energy Effluents. In *Energy Transition: Climate Action and Circularity* (pp. 169-267). American Chemical Society.
60. Hethnawi, A., Mosleh, A., & Nassar, N. N. (2021). Nanoparticles for Cleaning up Oil Sands Process-Affected Water. In *Nanoparticles: An Emerging Technology for Oil Production and Processing Applications* (pp. 445-496). Springer, Cham.

28c Low-Temperature Demulsification of High Waxy and High Water Cut Pickering Crude Oil Emulsion

Vidhya Vijaykumar, Jyoti Seth, Vinay Juvekar, Rochish Thaokar, Pooja Verma

Low-temperature demulsification of crude oil emulsion containing high concentrations of wax and water is rendered difficult due to Pickering stabilization by wax particles.

Secondary oil recovery methods are employed to extract oil as an oil well ages. This leads to an increase in the water content of the crude oil recovered from the well. Pickering emulsions are formed when the immiscible mixture of crude oil-water is transported from the well to the platform through the low-temperature sea bed, where the wax gets cooled well below the Wax Appearance Temperature. While precipitating out of the bulk crude, the wax associates with resins and asphaltenes, which are the polar fractions of the crude oil. The wax-resin-asphaltene composite exhibits amphiphilic behavior, enabling the wax composite particles to grow at the interface around the water droplets. This leads to the formation of wax-stabilized Pickering water-in-crude oil emulsions, which are difficult to destabilize. The extraction of crude oil from this stable emulsion is achievable only at high temperatures (greater than 353K). As a consequence of the increase in the water content of the emulsion, the heat capacity of the emulsion also increases. The limitation of available heating capacity in the heater-treater system, along with space constraints at the offshore platforms, necessitates that demulsification be carried out at lower temperatures, with high efficiency, in order to handle high water-cut crudes and realize high throughputs of de-emulsified crude oil. This demands chemical demulsifiers that can assist with the swift demulsification of these wax-stabilized Pickering emulsions, even at low temperatures. Low-temperature demulsification implies demulsification at temperatures close to the pour point of the emulsion and up to 10°C below it. High shear mixing may be required before settling, for temperatures much below the pour point.

From preliminary bottle test experiments, a specific combination of surfactants was found that could give 100% water separation within 5 minutes, even at low temperatures. The mechanism of destabilization, in the presence of these surfactants, is understood by studying the three-phase contact angle and interfacial rheology. The influence of surfactants on the three-phase contact angle between wax particles and the oil-water interface is understood using a customized setup comprising a camera and an infrared light source. Infrared light enables visibility through crude oil. The three-phase contact angle is determined by post-processing of the images using ImageJ software. Interfacial rheological analysis of the crude oil - brine interface is conducted using Du- Noüy Ring Apparatus (ARES-G2, TA Instruments).

The surfactants that act as effective demulsifiers wet the wax composite particles at the interface and displace them from the oil-water interface, thereby decreasing the three-phase contact angle to zero. Due to the displacement of the wax composite particles from the oil-water interface, the interfacial elasticity and yield strain reduces significantly. From the above understanding, we propose a guideline for the selection of surfactants for low-temperature demulsification of wax-stabilized Pickering crude oil emulsions, wherein the surfactant mixture must include a medium-HLB wetting agent in combination with a high-HLB phase inverting surfactant. The medium-HLB surfactants that act as good wetting agents are two-tailed, irrespective of whether ionic or non-ionic in nature. Due to their medium-HLB, they neither favor the oil nor the water phases and preferentially adhere to the interface. After the interface is devoid of the wax composite, a second high HLB surfactant assists in accelerated phase inversion of the water-in-oil emulsion to bi-continuous phase, which favors gravity settling of water. Multiple surfactant combinations, which were filtered based on the above -

mentioned criteria, have been tested in the oil field and have exhibited 100% separation efficiency within 5 minutes.

Controlled Release of Asphaltene Inhibitors from Nanomaterials in Oil and Gas Production

Authors:
Taeyoung Chang[1], Shuhao Liu[4], Chuxin Lei[2], Taehyung Park[1], Po-Wei Huang[3], Craig Klevan[3], Kurt Pennell[3], Linda Abriola[3], Guihua Yu[2], Keith P Johnston[1]
Mcketta Department of Chemical Engineering, The University of Texas at Austin[1]
Walker Department of Mechanical Engineering, The University of Texas at Austin[2]
School of Engineering, Brown University[3]
Department of Chemical Engineering, The University of Texas at Tyler[4]

Asphaltene precipitation significantly impacts the oil and gas industry by reducing production by depositing in pores upon decreases in pressure. Traditional preventive methods, such as squeeze treatment, require frequent and costly injections of asphaltene solvents or inhibitors (AIs) into production wells. To better maintain production, there is a substantial interest in developing asphaltene inhibitor-controlled release systems using nanoparticles to achieve long-term inhibition for up to one year. We propose an encapsulation strategy employing mesoporous silica nanoparticles and strategically designed AIs to facilitate the controlled, extended release of AIs in organic solvents including xylene and model oils.

We employed several characterization techniques, including scanning electron microscopy (SEM), nitrogen adsorption-desorption isotherms, dynamic light scattering, and zeta-potential measurements. The colloidally stable sub-500 nm nanoparticles could be injected with minimal formation damage in porous sandstones. Using UV-Vis spectroscopy, we assessed the inhibitor release behavior, to design chemical structures for prolonged release. A model was used to describe the kinetics and thermodynamics of the AI adsorption/desorption dynamics on silica to predict long term performance in reservoirs. By setting a new standard in asphaltene precipitation management, this approach not only aims to improve operational efficiency and reduce costs, but also to provide fundamental concepts for long term-controlled release with advanced nanomaterials in the oil and gas industry.

28e Effect of Normal Alkanes in the Solubility and CO_2 Thickening Effect of Poly-1-Decene Olygomer for IOR Application

Tomás Chávez-Miyauchi, Abbas Firoozabadi

Supercritical CO_2 ($scCO_2$) thickening is the most powerful approach to enable mobility control to increase sweep efficiency in improved oil recovery (IOR) processes As a consequence CO_2 can be stored in the subsurface in large quantities . Hydrocarbons in CO_2-rich phase may affect the performance of oligomers in CO_2 thickening for improved oil recovery (IOR). Hydrocarbons may also decrease the solubility of oligomers/polymers in CO_2. The 1-decene (P1D) oligomer with 20 repeat units shows excellent viscosification at high pressures and different temperature conditions. It is potentially an effective candidate for CO_2-IOR. In this report, a systematic study is presented on the solubility, and viscosification of CO_2 by P1D in the presence of n-hexane (nC_6), n-decane (nC_{10}), n-dodecane (nC_{12}) and n-hexadecane (C_{16}). Viscosity decreases as the concentration of hydrocarbon increases. The decrease is gradual and mild at concentrations of nC_6 and nC_{10} below 15 mol% in CO_2. P1D seems to be less affected by the presence of lighter hydrocarbons which would make it an effective candidate for IOR processes.

28f A Novel Core Annular Flow Inducing Device for Energy-Efficient Transportation of High Viscous Oil

Ekta Tayal, Sneh Singh, Sayan Halder, Gargi Das, Chirodeep Bakli, Subhabrata Ray

Introduction

Heading from the start of 21st century, the demand for crude oil has increased rapidly leading to a subsequent upsurge in its worldwide growth rate by 1.8%. The constantly increasing population and financial development have led to a major rise in the production of crude oil (over 80 million barrels per day), with an increasing proportion of heavy crude oil[1]. These oils exhibit high viscosity (>100cp) thereby requiring high pumping power for transportation[2]. Flow assurance during heavy crude transportation is also a challenge. To mitigate this, several cost-effective techniques have been adopted. One of them is transportation via the core annular flow (CAF) configuration. In this technique, the oil core is surrounded by an annular layer of water such that the pressure drop in the lubricated pipeline is almost close to the pressure drop required for flow of water only[3]. CAF also mandates the presence of a minimum inlet water fraction (conventionally termed as water cut) ranging between 10-30%. A higher water cut may lead to stable CAF, but does not necessarily ensure a decrease in pressure gradient. The increase in water cut additionally lowers the overall quantity of oil transported, also leading to separation issues.

Although core annular configuration has been known since long, it has not been widely applied in industries, a major challenge being to establish and stabilize CAF[3]. Past studies on oil-water flow patterns in pipelines have shown that core-annular flow is stable only for a narrow range of superficial oil and water velocities. Significant viscosity and density differences between the two fluids result in interfacial instabilities which disrupt CAF at even modest flow rates and cause stratification. This is particularly significant in horizontal pipelines, where buoyancy promotes oil core levitation[4]. A past study on the effect of bends and elbows have reported flow disruption due to centrifugal forces caused by change in the flow direction. The presence of fittings affects the oil-water interface, causing a reduction in energy saving[5]. The current state-of-the-art thus reveals that for commercial applications, CAF stability is a challenge except for density matched immiscible liquids. Re-establishment of core-annular flow downstream of fittings and maintaining the same is also an area which requires focus.

In the present study, a novel "CAF-inducing geometry" has been proposed to address these limitations. The purpose of the device is to ensure (i) establishment of stable CAF in horizontal pipelines, (ii) re-establishment of the disrupted flow after pipe fittings. The study also aims at forming a stable CAF with minimal water cut, ensuring a higher oil throughput. The effectiveness of the device is quantified from estimation of pressure drop and water cut, defined as the ratio of minimum water flowrate to the total flowrate necessary for stable CAF (Water cut = Water flowrate/ (Total oil and water flowrate)).

Experimental Setup

The experimental setup consists of a 3m long, 25mm diameter pipe (upstream section) with a gate valve positioned after 1m. The gate valve is selected as it is an inevitable fitting in pipelines. The "CAF-inducing geometry" is placed after the upstream pipe and is followed by an identical downstream pipe of 25mm diameter (Fig.1). Further, the gate valve causes flow disruption, and CAF is shown to be re-established following the proposed device.

The investigation is carried out using lubricating oil (density = 886.8 kg/m^3, viscosity 0.314 Pa.s) and tap water (density = 1000 kg/m^3, viscosity 0.001 Pa.s) as the test fluids. The superficial velocity of oil and water is kept within 1.02 m/s and 0.34 m/s respectively. Flow patterns at three sections, i.e., before valve (section I), between valve and "CAF inducing geometry" (section II), and after the "CAF inducing geometry" (section III) are recorded using a high-speed camera (Nikon, model Micro LC 320S). The pressure drop is measured by a differential pressure transducer (120STD type Honeywell Differential Pressure Transmitters with accuracy of ± 0.1%) and water cut is estimated from superficial velocities at the inception of CAF.

Prior to experiments, the gate valve is kept at 50% open position and the entire setup is filled with water. Subsequently, oil is introduced into the system at a given flowrate, and the water flowrate is decreased gradually to generate two-phase flow. The results are presented as flow pattern map and pressure gradient data. A few experiments are also performed with fully open gate valve and the efficacy of the device is established for both cases.

Results and Discussion

Flow pattern maps illustrate the observed flow patterns as a 2D plot with oil and water superficial velocity as the abscissa and ordinate respectively. Flow patterns in each of the three test sections is represented by three separate maps (Fig.2). Within the considered range of oil velocities (0.1 m/s - 1.0 m/s), the predominant flow patterns in Sections I and II are Stratified and CAF. However, beyond the valve (i.e., in Section II), most of the CAF transitions to stratified flow. In Section III, only CAF is observed.

Moreover, during experimentation two types of CAF are noted: thin and thick core annular flow. They are distinguished based on the thickness of the oil core (d) relative to the pipe diameter (D). The parameter $d/D>0.5$ indicates thick CAF, which is a preferred configuration for increased oil throughput and cost-effective oil-water separation.

A comparative analysis of flow patterns at the three sections reveals the following interesting features:

i. Gate valve disrupts CAF which is restored after the "CAF-inducing geometry".
ii. The distribution is thick CAF in Section III, while it is mostly thin CAF in section I, which implies the device effectiveness for higher oil throughput, regardless of the flow pattern upstream.
iii. Thick CAF in Section III is more stable at higher oil flowrates. At low oil flowrates, interface mixing to some extent is observed.
iv. The device facilitates a higher oil throughput, with a significantly lower water cut (6.25%) compared to section I & II where it is 16.7%.

Given that the primary objective of transporting viscous oil via the CAF configuration is to minimize pressure drop, the pressure gradient in both the sections II & III, are reported in terms of a pressure reduction factor (ratio of the pressure gradient for two-phase flow to pressure gradient for single-phase flow of oil only). The measurements show that at the higher velocity of 1 m/s, pressure reduction factor at Section III is 0.127. This improvement ensures an energy-efficient transportation with an energy saving of 87%.

Conclusion

In summary, we propose a device that can establish CAF at its downstream, irrespective of the upstream flow pattern during high viscous oil-water flow. The effectiveness of the proposed device is demonstrated for a partially opened (~50%) gate valve. The findings indicate that within the range of oil velocities (0.1 m/s - 1.0 m/s), the presence of gate valve exacerbates oil stratification. However, upon encountering the device, thick core annular flow forms in the downstream section. Notably, the minimum water cut for stable CAF is 16.7 % both before and after valve, while it is 6.25% downstream of CAF inducing geometry. Therefore, the device not only generates CAF at high throughput with a lower water cut, but provides an energy saving of 87% at an oil flowrate of 30 Lpm.

References

[1] D. D. Joseph, R. Bai, and K. P. Chen, "Core-Annular Flows," pp. 65–90, 1997.

[2] R. Martínez-Palou *et al.*, "Transportation of heavy and extra-heavy crude oil by pipeline: A review," *J.*

Pet. Sci. Eng., vol. 75, no. 3–4, pp. 274–282, 2011, doi: 10.1016/j.petrol.2010.11.020.

[3] A. Hart, "A review of technologies for transporting heavy crude oil and bitumen via pipelines," *J. Pet. Explor. Prod. Technol.*, vol. 4, no. 3, pp. 327–336, 2014, doi: 10.1007/s13202-013-0086-6.

[4] N. M. de A. Coelho *et al.*, "Energy savings on heavy oil transportation through core annular flow pattern: An experimental approach," *Int. J. Multiph. Flow*, vol. 122, p. 103127, 2020, doi: 10.1016/j.ijmultiphaseflow.2019.103127.

[5] F. Jiang, J. Chang, H. Huang, and J. Huang, "A Study of the Interface Fluctuation and Energy Saving of Oil–Water Annular Flow," *Energies*, vol. 15, no. 6, 2022, doi: 10.3390/en15062123.

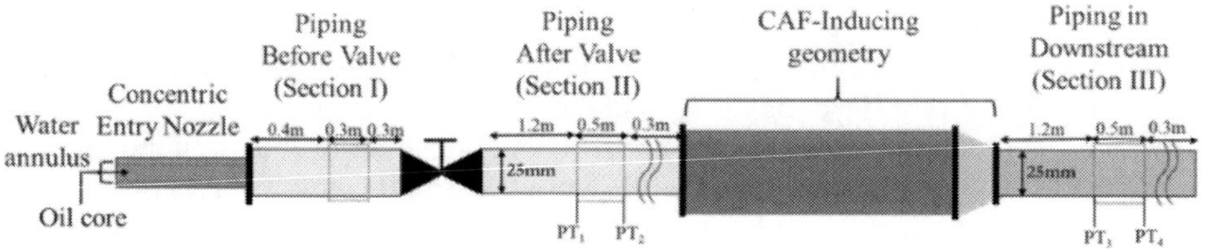

Fig. 1: Schematic representation of the test rig with a gate valve (50% open)

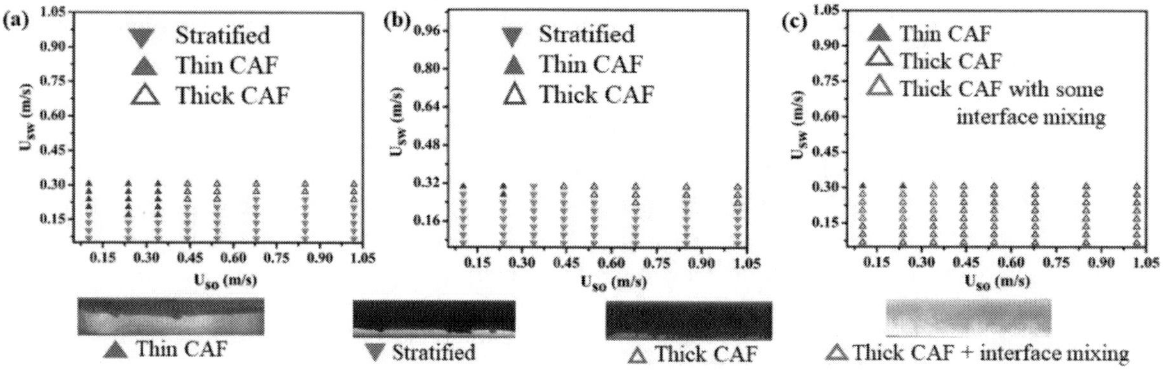

Fig. 2: Flow pattern map for (a) Section I (b) Section II (c) Section III

28g Paraffin Deposition in Stratified Gas and Oil Production Lines

Abdulaziz Alhejaili, Elijah Bell, Nagu Daraboina

As the global energy demand continues to increase, the oil and gas industry has had to expand its operation to ultradeep water and extreme environments. This has introduced many challenges to flow assurance, making ensuring products' profitable and safe flow a greater challenge. Wax deposition is one such major challenge to flow assurance, as wax crystals are prone to precipitate and deposit as the oil temperature drops below the wax appearance temperature. Understanding this phenomenon and how to prevent or mitigate it is crucial to minimize associated risks and losses to production. While most of the work in this area is focused on single (liquid) phase conditions, wax deposition could occur in multiphase as well. The difference in hydrodynamic and heat transfer in multiphase and their dependence on flow regime could affect the wax deposition and mitigation strategies.

This work investigates wax deposition in stratified gas and oil flow in pipes. It was found that, while in stratified smooth deposition occurred at the bottom section of the pipe only, as the flow transitioned to stratified wavy, deposition in the top was observed due to liquid contact with the top section of the pipe. Here, three stages to explain the deposition in the stratified flow regime were observed. Stage 1 represents the initial deposition at the bottom of the pipe due to direct contact between the cold pipe wall and the liquid phase. Stage 2 starts as the deposition starts to appear at the upper portion of the pipe wall. Lastly, stage 3 defines the slowing in wax deposition growth and increase in hardness as the complete pipe wall is covered. These findings will aid in developing better prediction models and establishing guidelines to minimize the effect of wax deposition issues in multiphase flow.

Towards an Integrated Wide Approach for Sustainable Upstream Field Recovery

Shakeel Ramjanee,[a] Professor Nilay Shah[a]

[a] *Department of Chemical Engineering, Sargent Centre of Process Systems Engineering, Imperial College London, South Kensington, London, SW7 2AZ, UK.*
n.shah@imperial.ac.uk

Abstract

Integrated asset modelling (also known as integrated production modelling) is the modelling and simulation of an entire production facility consisting of both subsurface and surface elements. The intent is to holistically capture the complex interactions between the individual components of the system. Historically, asset modelling has entailed discrete reservoir and facility models; this mindset is however slowly altering towards integrated asset modelling for production forecasts and optimisation whereby both system-imposed constraints i.e. fluid flow from the reservoir and surface handling (i.e. capacity) constraints, can be considered simultaneously. Adopting an integrated approach provides a holistic overview to modelling the production system which enables the user to not only understand the system constraints but also determine where optimisation of field parameters could be beneficial for improving production. The full benefits of these models however are not realised for several reasons, including extensive simulation time of complex models, siloed disciplinary approaches resulting in separate disciplines owning separate elements of the framework.

This research study focuses on the development of surrogate models of an integrated field set up for both a gas producing and oil producing supplemented by water injection field. Surrogate models were developed via several techniques and subsequently tested to determine the accuracy versus the fully integrated asset industry model. Global sensitivity analyses were performed on a selected number of input (manipulated) field variables on the overall outputs from the surrogate models to validate the surrogate relationship postulations prior to attempting optimisation of the developed surrogates with varied results. Following this, constrained non-linear and gradient free optimisation of the derived proxy models was carried out to determine whether these could be used to predict the optimum production and corresponding pressure profiles. In summary, the surrogate models developed via traditional methods (e.g. response surface models and manual regression) and via more complex methods (e.g. Artificial Neural Networks, ANN and Random Sampling High Dimensional Modelling Representation RS-HDMR) provided appropriately accurate predictions when compared to a commonly deployed fully developed integrated asset model from Petroleum Experts within the upstream industry. Deviations from the predicted profiles in comparison to the integrated asset model outputs were noted at later field life or in circumstances when production is significantly choked back which would conclude that the surrogate models developed are more suited to be used in certain phases (early) of a field development or life cycle of a field hence within an opportunity realization framework, within the assess, select concept framing stages. Instabilities within the predictions of the surrogates were noted and attempt was made to smooth these out to improve the surrogate model definition by deploying smoothing splines. Lastly, the offset between the full physics reservoir coupled integrated model and the simplified integrated model was assessed and optimised with view to utilising the reduced order model to predict the performance of a full physics integrated asset model. The developed surrogate models can be deployed to optimise and forecast production in a fraction of the simulation time; given the forecasted demand of hydrocarbons within the energy mix to meet the global energy demand in the short to medium term; optimising recovery from existing fields is critical especially given the reduced capex investments into new i.e. greenfield fields being developed.

Keywords: Integrated Asset Modelling; Optimisation; Surrogate model development; Reservoir modelling; Sustainable recovery

1. Methodology

Upon setting out on this research post review of the available literature which concluded that prime focus is on reservoir rather than integrated model simplification, attempting to construct an integrated asset model test case to use a basis and develop further understanding of these tools was deemed to be the first step. The following section outlines the industry tools assessed, the subsequent test cases built and the outputs from the simulation forecasts. These models would then be used as a basis for investigating surrogate model development methods. the most widely applied integrated asset modelling tool within the upstream industry was concluded to be the Petroleum Experts (PETEX ®) software suite; as such, this was selected as the basis for the research following application and receipt of education licenses at Imperial College London; an Intel i7 core CPU of 3.4GHZ with 8GB RAM workstation was used throughout this research.

1.1. Development of a Case Study Model within PETEX® Framework

Initially, a fully integrated field model (based on a central North Sea field) was developed within the industry state of the art tool, PETEX® integrated asset modelling suite to be utilised as a starting case for development of a proxy model representation. The system consisted of 3 black oil reservoir tanks within MBAL (PETEX® Material Balance zero-dimensional reservoir modelling tool), 8 producing wells across 2 production manifold, 3 production flow lines commingling at a landing host facility separator operating at 20bar. Following the construction of the test case within PETEX®, a simplified approach was adopted as a starting point for evaluating an integrated asset model proxy model proposal. The driver for the simplified approach was to derive a model that could be used as a framework for further complex cases in future; two cases for a gas only approach and oil and water injection approach were selected to cover both a primary (natural pressure support) and secondary drive mechanism (additional pressure support). As such, a single gas well coupled to a single flowline hooked up to a single host receiving facility was developed in GAP (PETEX® General Allocation Package - surface modeller within PETEX® suite). A standard tank reservoir model within Mbal was coupled to a single gas well within GAP. This framework was subsequently extended to an oil producing with water injection support field case.

1.2. Global Sensitivity Analysis within GSA Sobol and Generation of RS-HDMR Proxy Meta Model Representations

Utilising the Sobol GSA Proxy meta modelling software (developed internally by Imperial College London specifically for global sensitivity evaluation), a global sensitivity analysis was carried out utilising the input and output data extracted for 164 single solve simulation runs within GAP. The aim was to identify which of the selected

3 manipulated variables had the highest impact on the output (gas production). The input and output data structure was set up as follows:
Input parameters:

- Input parameter 1: Separator landing Pressure (barg)
- Input parameter 2: Roughness ratio of flowline (dimensionless)
- Input parameter 3: Gas well choke differential pressure (barg)

Output parameter:

- Output parameter 1: Optimised Gas production (MMScfd)

Within Sobol GSA, the input distribution was generated by specifying a lower and upper bound with the distribution type for each parameter (selected as uniform for each of the three input parameters); the lower and upper bounds were selected based on the realistic magnitude of each parameter in addition to capturing the physical constraints of the system e.g. excessively high differential pressure across the well choke will result in no flow from the well. This generated a random data set which was then used as specified inputs within the GAP model to generate an optimised (single solve) gas production forecast (MMscfd). Utilising this input and subsequent output generated data set, a sensitivity analysis was then run within GSA Sobol utilising the RS-HDMR method (Li et al,2002) of analysis with a Sobol sequence sampling strategy.

1.3 Development of a Gas Well Integrated Asset Model Incorporating a Coupled Numerical Reservoir Model (Eclipse® and Mbal)

Following the successful development of a test integrated field single well gas within GAP and Resolve (PETEX® integrator module to couple reservoir models and surface/well models), the subsequent step of the research was to attempt to develop a numerical reservoir gas model to couple to the integrated asset model via Resolve (PETEX® integrator tool) The aim was to firstly incorporate a detailed reservoir representation to compare simulation time and results versus the simplified integrated tank reservoir model. Secondly, the aim was to extend the research to a more representative reservoir model that is typically employed by subsurface teams within the upstream industry. Two full physics gas only reservoir models were developed in Eclipse® (3D reservoir modelling tool):

- Single well gas reservoir model without aquifer support
- Single well gas reservoir model with numerical aquifer support incorporating water breakthrough after 'X' years of production.

The developed 3D reservoir (full physics) models were subsequently coupled to a network GAP model consisting of a single flowline and arrival separator via Resolve. The coupled network was then used as a basis for further proxy modelling development and sensitivity analyses. In addition, a simplified tank reservoir model with a single gas well within Mbal (PETEX®) was developed and coupled to the integrated asset model via Resolve; this was used as a comparison to compare the output and performance of the full physics reservoir model against the simplified tank model.

1.4 Development of an Oil and Water Injection Integrated Asset Model Incorporating a Coupled Numerical 3D Reservoir Model

The Norne reservoir model (NTNU) representing a segment of the Norne field was loaded into Eclipse®. The Eclipse® based Norne reservoir model was successfully run as a standalone simulation via the Schlumberger simulation platform. The subsequent step was to couple the reservoir model to an integrated asset framework utilising the PETEX® simulation suite. Within GAP, a network model was developed comprising of 3 producing wells, a common production flowline and a production separator set an arrival pressure of 22 barg. On the recommendation from PETEX®, the water injection network comprising of 2 water injection wells and a common water injection manifold was configured in a separate GAP network model. Both GAP models were then coupled to the Norne Eclipse® Reservoir simulation deck via the PETEX® Resolve platform. Following the successful coupling of the Eclipse® model to both the production and injection GAP models within Resolve, the network model was subsequently test run for several forecast cases. Initially the water injection choke position (Xw) on both injection wells was varied for a fixed oil producer choke pressure differential (Xo) setting. For comparison, the water injection choke position (Xw) was then fixed at 250bar (38% choke position) equivalent and the oil producing well pressure differential was varied (Xo). The forecast runs were specified from the start of field life set at November 1997 for a production life of 20 years across a 3-day time step.

1.5 Sensitivity Analysis Executed via GSA Sobol and Subsequent Development of RS-HDMR Proxy Model Representations Incorporating Time Dependency

The GSA Sobol (Imperial College London) sensitivity analysis software applied to the gas well case was deployed for the oil/water injection model case and used to analyse the key manipulated variables to determine which input parameter displays the most impact on the output parameters. The input and output data sets were based on the single solve optimisation results in GAP where the simulator forecasts the optimum oil and gas production respectively for the user specified inputs and hence is effectively considered to be time independent.

The input and output parameters for the sensitivity analysis was configured as follows:

Input parameters:

- Input Parameter 1: Production choke setting, Xo (%)
- Input Parameter 2: Water injection choke setting, Xw (%)
- Input Parameter 3: Separator landing pressure, P (barg)

Output parameters:

- Output Parameter 1: Oil production (bbl/d)
- Output Parameter 2: Gas production (MMscfd)

164 single solve simulation runs were executed in a batch manner in GAP to generate the input and output data sets. In the absence of any other information and in line with the maximum entropy principle, the input parameter distribution was specified as uniform, extracted, and run within the integrated asset model to generate the corresponding gas and oil production outputs. From the subsequent sensivity analysis, the water injection choke setting, Xw (Input Parameter 2) has the least impact on the output with the production choke, Xo having the highest impact. Following the generation of the time independent RS- HDMR meta model within GSA Sobol, an extension to apply this to a

time dependent meta model was investigated. The driver for this was to factor in that the production at different time steps varies particularly when comparing early field life to late field life (short to medium- and long-term decision horizons). The simulation runs required at least 10 minutes per forecast run within PETEX® hence required considerable simulation time to generate the full forecasted data set. GSA Sobol generates an output for each individual time step i.e. for each of the 215 time steps, a sensitivity per output parameter is reported for each of the three input parameters.

1.6 Generation of Artificial Neural Network (ANN) Proxy Model Representation

Following the successful coupling of the reservoir models (both material balance within Mbal and numerical models developed within Eclipse®), to the PETEX® GAP network via Resolve, these were then utilised as a template to develop further surrogate representations of the two selected key parameters, the change in reservoir pressure with respect to time and the gas production output as a function of key specified input parameters was postulated per Eq(1) and Eq (2):

$$dP/dt = f(G(t)) \quad (1)$$

dP/dt references the change in field pressure with respect to time and G(t) is the gas production

$$G(t) = f(dP,Xg) \quad (2)$$

dP references the differential between the field pressure and landing i.e. separator pressure (barg) and Xg references the choke setpoint on the producing gas well (%). An Artificial Neural Network (ANN) surrogate model utilising the machine learning toolbox within Matlab was generated utilising the data set for both the integrated tank reservoir model (with acquifer support) as well as the results obtained from the coupled material balance reservoir model to the integrated asset model within PETEX® . The ANN fit was applied to both dP/dT f (G(t) and G(t) f (Xg, delta P) data sets for both tank and Eclipse® reservoir model cases. The data set was imported as fixed arrays within Matlab and the Neural Fitting Tool was employed to train the neural network model The generated ANN model within Matlab was subsequently tested to generate a set of outputs for a given specified array of inputs and compared to the corresponding output array within PETEX® . In summary a positive R^2 correlation (0.99) was observed between the training and test data set for the tank reservoir model case. A training algorithm of Baysian Regularization was selected based on its applicability for noisy and difficult datasets with typically 10 hidden neurons specified; specifying more hidden neurons resulted in a poorer fit and ultimately longer training times. Positive R2 values (0.99) were further confirmed by calling on Matlab to evaluate the given outputs i.e. dP and X for a given Gas production rate, G(t) and evaluate dP/dT for a given gas production rate, G(t).

1.7 Refining of Differential Pressure Profile Utilising Smoothing Spline Fit (Gas Field and Oil Field supplemented by Water Injection Cases)

Following the attempted development of the surrogate representation of both G(t) and dP/dT functions, the results obtained for the dP/dT profile were suboptimal in comparison given the R^2 values obtained for the fit of dP/dT versus G(t). On analysis of the differential pressure profile extracted from the PETEX® simulation forecast runs within Resolve, it was evident that there were significant oscillations within the data set that may have contributed to the sub optimal fit of the polynomial per the response surface modelling method. Based on this, a smoothing spline approach was attempted to smooth out the oscillations. This data set was extracted and loaded in Matlab to generate a similar curve fit as well as utilising the third-party surrogate modelling tool, ALAMO (The Optimization firm LLC) to generate a surrogate representation between dPf/dt and G(t). ALAMO is a machine learning software utilised for building models from data collected and generate algebraic surrogate models. Based on the revised spline differential estimate, ALAMO was used to generate a functional relationship between dPf/dt and G(t). Utilising the generated spline fit differential estimates and the extracted data profiles, an Artificial Neural Network (ANN) model was also derived for both functional relationships applied to the oil and water injection case:

O(t) f (Xw, dP) **(3)**

O(t) represents the oil production output, dP represents the change in field pressure less landing pressure (barg), and Xw the choke setpoint on the water injection well (%)

dP/dt f (Xw, O(t)) **(4)**

dP/dt is the change in field pressure with respect to time and Xw the choke setpoint on the water injection well (%); later extended to dPb/dt as a sensitivity denoting change in bottom hole pressure.

2.0 Results

The following section aims to summarise the key results from the conducted research at Imperial College London. The sensivity analysis carried out with GSA Sobol (Imperial College London) for the gas well integrated asset model generated the following sensitivity analysis output per Figure 1 which displays Si, the first order sensitivity of each input parameter towards the output parameter; the higher the Si value i.e. closer to 1, the higher sensitivity towards the output. The separator landing pressure i.e. input parameter 1 demonstrated the largest sensitivity towards the gas production output.

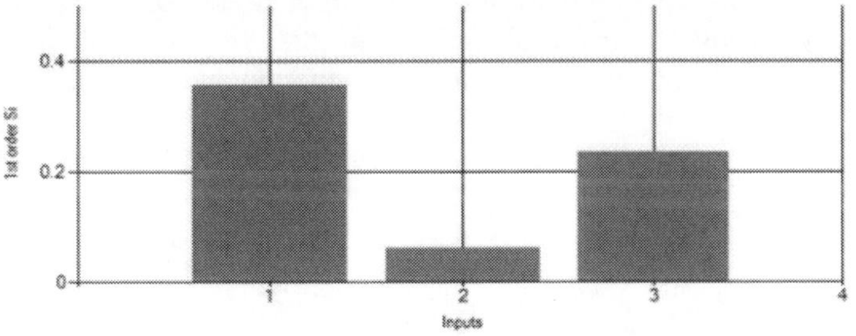

Figure 1: Sensitivity Analysis of input parameters within GSA Sobol; Si (Sensitivity Index) vs Input Parameters

The subsequent RS-HDMR (Random Sampling -High Dimensional Modelling Representation) meta model generated within Matlab was tested and this generated a positive R^2 correlations at 0.90. The HDMR model itself is deployed via Matlab to provide an accurate prediction of the production and field pressure profile; incorporation of time dependency demonstrated positive R^2 correlations at early field life conditions but deteriorated at later field life. Figures 2-5 depicts the production and pressure profile predictions from the most optimal surrogate models developed (Response Surface Method, ALAMO, ANN, RS-HDMR methods were deployed with varying degrees of success).

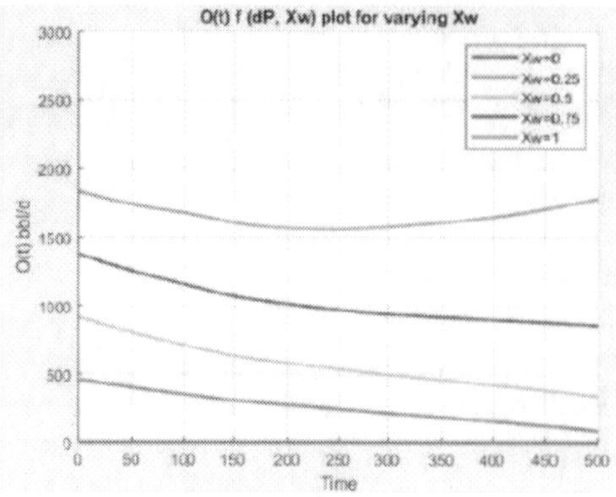

Figure 2: Oil production prediction via surrogate model at varying water injection choke set points, Xw

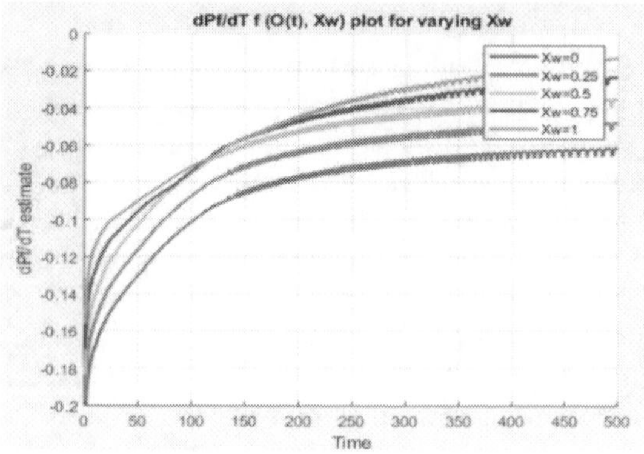

Figure 3: Field differential pressure prediction (oil field) via surrogate model at varying water injection choke setpoints, Xw

A constrained non-linear optimisation framework (Fmincon) based on the developed surrogate relationship for gas production and differential pressure profiles was set up utilising a custom ODE solver (adapted for stiff differential equations). The resultant cumulative gas production output compared favourably with the corresponding cumulative production output from the integrated asset simulator model. By specifying a time step, an initial guess for the input parameters, the framework can maximise G(t) over a given time frame. Further optimisation of the developed surrogate models (in addition to optimising the differential i.e. epsilon, ε, between the full physics reservoir coupled integrated asset model and the simplified tank reservoir integrated asset model) was investigated. Optimising the RS-HDMR and ANN models utilising metaheuristic methods was attempted; genetic algorithms were employed to optimise the developed ANN's. Sobol Opt (GSA Sobol global optimiser which is designed to solve global non-linear optimisation problems was deployed to optimise the RS-HDMR models which resulted in slight overpredictions in cumulative production and revenue.

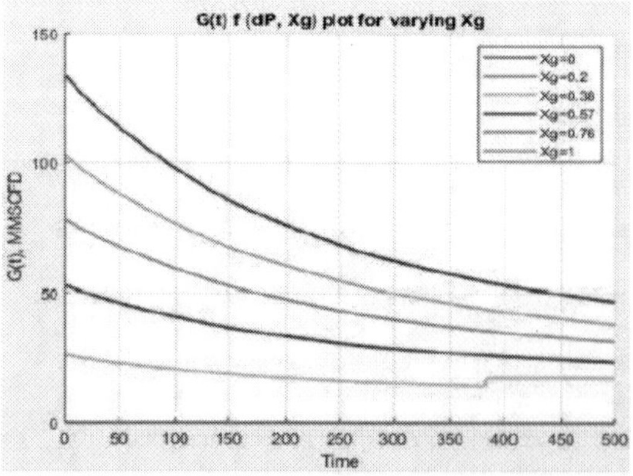

Figure 4: Gas Production prediction via surrogate model at varying water injection choke set points, Xw

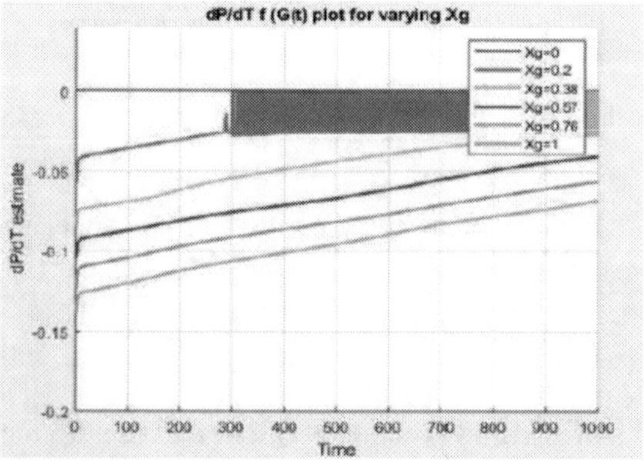

Figure 5: Field differential pressure predictions (gas field) via surrogate model at varying gas well choke setpoints, Xg

3.0 Conclusions

Development of surrogates was attempted by utilising a number of integrated asset model frameworks (simplified gas tank model including acquifer support, full physics gas tank model and full physics reservoir simulator). Having developed the integrated asset model test cases within the industry state of the art PETEX® suite and successfully run several global sensitivity assessments, several approaches were investigated to test fit a relationship for G(t), O(t) and the field/bottom hole pressure profiles, dPf/dt and dPb/dt. Response surface fitting of polynomials demonstrated good correlation for the production functions but less so for the field pressure profiles. This was the case for both gas and oil production integrated asset model surrogate development attempts. Imperial College London's proprietary meta modelling software, Sobol GSA was deployed to determine whether extracted input and output datasets could be utilised to generate an RS-HDMR metamodel for predicting absolute production and pressure profiles for a given specified set of input parameters. Employing Artificial Neural Networks (ANN) demonstrated a positive response when training the data sets utilising the Bayesian training regime; the subsequent developed ANN's can be used to predict the production and the pressure profiles for a given specified user set of inputs i.e. Xg, dP and Xw. Optimisation of the developed surrogates was attempted which was successful for a constrained non-linear optimisation framework, slight under predictions were noted when optimising via gradient free methods such as genetic algorithms applied to developed Artificial Neural Networks. Lastly, machine learning was adopted to attempt to adapt the reduced order integrated asset model network to predict the performance of the full physics integrated asset model with similar findings i.e. deviations noted in later field life conditions which would restrict the usage of the proxy models at these conditions.

References

[1] ALAMO (Automated Learning for Algebraic Models for Optimisation) Modelling Tool, The Optimization Firm, LLC, https://minlp.com/ALAMO-modelling-tool (Software); downloaded at Imperial College London in December 2020

[2] Li, G., Wang, S., Rabitz, H., (2002), Practical Approaches To Construct RS-HDMR Component Functions, Journal of Physical Chemistry A 106, 8721-8733

[3] NTNU, Norne Reservoir Database, Department of Geoscience and Petroleum, Norne, http://www.ipt.ntnu.no/~norne/wiki1/doku.php?id=english:license,accessed at Imperial College London in September 2018

[4] SobolGSA, Imperial College London, Department of Chemical Engineering, Centre for Process Systems Engineering, London, SW72AZ, UK (Software accessed internally at Imperial College London), Downloaded January 2018

28i Machine Learning on Heterogeneous Data Sets of Adsorption Energies

Avery Hill, Gbolade Kayode, Matthew Montemore

Machine learning models have been developed that can accurately predict adsorption energies for use in catalyst screening, but these models are not well-suited for fitting to heterogenous data sets containing data with different computational setups and on different sets of materials. More broadly, existing models often perform poorly when they encounter data that is different from their training environment. Often, these models fail to implement chemical knowledge that would aid in prediction accuracy and make their results more interpretable, such as the fact that elements are discrete entities. Here, we develop and apply a framework that can effectively learn from a heterogeneous dataset, allowing us to simultaneously leverage multiple existing datasets and allowing very efficient transfer learning to new applications.

To perform this fitting on heterogeneous datasets, we develop a multilevel machine learning model that treats each element as a discretized entity, and uses latent variables to link them to adsorption energies. Essentially, the framework learns how the local environment around a given element modifies its chemical properties towards a broad set of adsorption energies. Our model has been shown to make accurate binding energy predictions on heterogeneous data and enables efficient transfer learning, allowing researchers to efficiently use existing datasets to accelerate screening with their own computational setup on their own problem of interest.

28j CO_2-Soluble Wettability-Altering Surfactants for CO_2 Huff-and-Puff in Fractured Formations

Abdullah Shaheer, Lauren Burrows, Angela Goodman, Deepak Tapriyal, Foad Haeri, Robert M. Enick

Despite the prolific increase in oil production due to the use of horizontal fractured wells in formations such as the Bakken (NK) and Eagle Ford (TX), only 5% of the oil in these formations is produced during primary (natural pressure-induced) recovery. To recover some of the remaining 95% of the oil via CO_2 "huff-and-puff", high pressure CO_2 is injected into the fractured shale. The CO_2 is allowed to diffuse into the oil retained within the very low permeability shale for a while, and then a CO_2-oil mixture flows back to the surface as the well is opened for production. CO_2-induced oil recovery is attributed to several mechanisms, including oil swelling, oil viscosity reduction, extraction of oil into CO_2, and dissolution of CO_2 into oil. However, significant wettability alteration is not a notable mechanism for CO_2 huff and puff enhanced oil recovery.

CO_2 huff-and-puff results in the permanent storage of a portion of the CO_2 in the shale, the production of oil to the surface, and the simultaneous production of some of the CO_2 that can be captured and re-used in the next huff and puff cycle.

The focus of this research project is related to improving CO_2 huff-and-puff via the dissolution of a dilute concentration (0.01-1.00 wt%) of an additive in CO_2 that changes the fractured shale wettability from oil-wet toward CO_2-wet. Shale wettability alteration from oil-wet toward CO_2-wet occurs due to surfactant adsorption and/or oil emulsification. (There are no commercially available CO_2-soluble cationic surfactant candidates that could alter wettability via removing naturally occurring anionic oil components via ion pairing.) This concept, if successful, will impart another mechanism to the CO_2 huff and puff process that enhances oil recovery.

Each wettability-altering candidate is a commercially available compound that contains a CO_2-phile (e.g. carbonyl, polyethylene oxide or propylene oxide) and an oil-phile (e.g. alkyl group). The solubility of surfactants in CO_2 was determined as a function of temperature and pressure. The surfactant candidates included an ethoxylated tridecyl alcohol, a propoxylated tridecyl alcohol, 3-pentanone, Span 20, and Tween 20. CO_2 cloud points curves (i.e. solubility curves) for these compounds were determined, and those surfactants that were soluble in CO_2 at typical CO_2 huff and puff conditions were tested for wettability alteration.

Wettability alteration measurements consisted of high pressure contact angle experiments in which flat oil-wet shale samples were immersed in high pressure CO_2-surfactant solutions, and a drop of oil was placed on the shale surface. In the absence of surfactant, the oil droplet spread on the oil-wet shale. However, when surfactants were present in the CO_2, shifts in wettability from strongly oil-wet toward intermediate oil-CO_2 wet or CO_2-wet were observed as the oil droplet beaded up on the shale.

CO_2-oil interfacial tension (IFT) at high temperature and pressure is affected by the surfactant in CO_2 As shown in several recent literature reviews, reduction of CO_2-oil IFT to ultralow values is not necessary for this process; oil recovery via wettability alteration is actually promoted by significant changes in wettability coupled with modest reductions in CO_2-oil IFT.

Finally, CO_2 huff-n-puff experiments were conducted with CO_2 and CO_2-surfactant solutions using small cylindrical shale cores. The shale core was first aged in a high-pressure hot oil for a week to ensure the shale is oil-wet and oil-saturated. The shale was weighed before it was immersed in a high-pressure solution of CO_2 or CO_2-surfactant solution for 24 hours. The vessel was then vented over a 10-minute period, during which time oil and CO_2 emerged from the shale. Oil recovery was determined via material balance. In-situ NMR measurements will also be used to determine the distribution of oil and CO_2 and oil within the fractured shale during and after the experiment.

A rankling of the wettability-altering agents indicates that some of the candidate surfactants have little discernible impact on oil recovery, however others induced an increase in the fraction of oil recovered during the huff and puff process that could justify the increased cost of adding surfactant to the CO_2.

28k Formulation and Evaluation of Nanofluids for Enhanced Oil Recovery in Tight Carbonate Reservoirs: A Laboratory Study

Fatemah Al-Shareedah, Osamah Alomair, Abdulaziz Al-Dhwaihi, Nyeso Azubuike

Oil holds significant global importance as a vital energy resource, accounting for more than 30% of the total primary energy consumption globally. Oil production from conventional oil reservoirs is declining amidst a growing global energy demand. Unconventional oil reservoirs, such as tight carbonate rocks, are significant hydrocarbon reservoirs containing approximately 60% of the world's oil reserves. However, their characteristics, including natural fractures, unfavourable wettability, low permeability, and heterogeneity of rock properties, result in typically low oil recovery using primary and secondary recovery mechanisms. This study investigated the combined effects of formation water with 213,000 ppm salinity, cationic surfactants at a critical micelle concentration of 0.25 vol.%, and three distinct nanoparticle types (SiO_2, γ-Al_2O_3, and Al_2SiO_5) on altering carbonate rock wettability and the resultant enhancement in oil recovery. To assess the surface effects and wetting behaviour of the rock when interacting with various injection fluids, fluid-fluid interfacial tension (IFT) and fluid-rock contact angle tests were performed. The oil displacement efficacy of each injection fluid was investigated through dynamic core flooding experiments, conducted at a constant injection rate of 0.5 cc/min at 75°C. The results indicate that the IFT and contact angle for the formation water are 6.27 mN/m and 84.1°, respectively. However, the lowest IFT and most significant alteration of rock wettability towards water-wetness were observed when 0.25 vol.% surfactant was added to 0.075 wt.% of the respective nanosuspension. The recorded IFT and contact angles for Al_2SiO_5, Al_2O_3 and SiO_2 nanosuspensions are 1.23 mN/m, 67.5°; 2.18 mN/m, 59.1°; and 3.01 mN/m, 60.2°, respectively. The dynamic displacement results showed that injecting 2 pore volumes (2PV) of formation water achieved 38.69% oil recovery of original oil in place (OOIP). Surfactant flooding at 0.25 vol.% increased the oil recovery by 13.71% at 2PV compared to the formation water. Oil recovery was highest at a nanoparticle concentration of 0.075 wt.% for each type tested, with Al_2SiO_5 yielding the best results, followed by γ-Al_2O_3, and SiO_2. The combination of 0.075 wt.% nanoparticles with 0.25 vol.% surfactant further improved recovery, ranking the nanoparticles effectiveness as γ-Al_2O_3> SiO_2 > Al_2SiO_5 based on the surfactant's impact. The optimal strategy for maximized oil recovery was flooding with 0.075 wt.% Al_2SiO_5 concentration and 0.25 vol% surfactant concentration, leading to an increase to 63.05% at 1PV and approximately 80.31% after 2PV injection, demonstrating the efficacy of composite nanoparticles combined with surfactants in enhanced oil recovery from tight carbonate reservoirs.

Keywords: Carbonate reservoirs, enhanced oil recovery, surfactants flooding, nanoparticles, waterflooding, wettability alteration, interfacial tension, contact angle.

28l Evaluation of Locally Developed Zwitterionic Surfactants for Enhanced Oil Recovery Application in Harsh Reservoirs

Muhammad Israr, Masooma Nazar, Syed Muhammad Shakil Hussain, Muhammad Shahzad Kamal, Mohammad Alotaibi, Mohanad Fahmi

In response to the increasing demand for efficient and sustainable energy solutions, the oil and gas industry endures to face the challenge of maximizing productivity from reservoirs under harsh environmental conditions, such as high-temperature and high-salinity. Chemical enhanced oil recovery (cEOR) methods play a crucial role in this endeavor, offering promising routes for improving oil production. Nevertheless, conventional EOR chemicals frequently encounter stability issues and significant adsorption challenges.

This work presents pioneering research on the development and utilization of locally developed betaine based polyoxyethylene zwitterionic surfactants for enhanced oil recovery (EOR) in high-temperature and high-salinity reservoirs, concentrating on their technical efficiency and practical applications. Our study, addresses the molecular design and characterization of zwitterionic surfactants tailored to endure high temperatures, and salinities, found in challenging reservoirs. Spectroscopic techniques, including nuclear magnetic resonance (NMR), Fourier transform infrared (FTIR), and Thermogravimetric analysis (TGA) were used to confirm the chemical structure and thermal stability of the surfactants, respectively.

Thermogravimetric Analysis (TGA) data provide compelling evidence that the zwitterionic surfactants exhibit excellent thermal stability, demonstrating exceptional resistance even under elevated temperatures of up to 350°C. Systematic laboratory analysis, involving interfacial tension studies, and wettability alteration have determined that these locally produced surfactants are quite efficient in increasing oil recovery efficiency in harsh reservoirs. Moreover, compatibility studies with high and low salinity water (Formation water, Sea water, Deionized water) demonstrated the surfactants' robust performance and practicable applicability in Middle East reservoirs.

This work provides an important advancement in EOR technology, proposing a promising option to enhance oil recovery while conquering the challenges faced by harsh environmental conditions.

476b Electro-Biodiesel Empowered By Co-Design of Microorganism and Electrocatalysis

Kainan Chen, Peng Zhang, Yayun Chen, Chengcheng Fei, Jiali Yu, Jiahong Zhou, Susie Dai, Joshua Yuan

Developing an efficient and sustainable method for fuel production is essential for reducing reliance on fossil fuels and advancing humanity's sustainable development goals. Proposed solutions like biofuels or electro-fuels seek to address this issue, but encounter constraints in either efficiency or the carbon chain length of their products, hindering their efforts to decarbonize critical sectors like long-range heavy-duty vehicle fuels. We have integrated electrocatalytic CO_2 reduction reaction (CO_2RR) for biocompatible C2+ production with microbial fermentation to generate lipids, establishing a novel route for CO_2-to-electro-biodiesel production. For the first time, we unveiled metabolic constraints in microbial C2 utilization through simulations and metabolomics. This systems biology analysis guided the further design of synthetic biology approaches to attain reductant balance, boost ATP production, enhance lipid conversion efficiency, and increase lipid yield. Additionally, we identified optimal ratios of ethanol and acetate to achieve co-substrate synergy for lipid production. Moreover, we designed a bimetallic catalyst tailored to produce the desired C2 product profile, facilitating an optimal lipid conversion. The co-design of microorganism and catalyst has enabled the electro-biodiesel system to achieve a solar-energy-to-lipid conversion efficiency of 4.6%. This platform combines efficient CO_2RR that can be powered by renewable energy with lipid biosynthesis using microbial cell factories, potentially reducing land use by 48-fold compared to current biodiesel methods and potentially leading to negative carbon emissions in contrast to traditional gasoline.

476c Development of a Soft Sensor for Intensified Ethanol Fermentation Process Using Deep Learning and Augmented Dataset

Hyun Kwon, Elmer Ccopa Rivera, Celina K. Yamakawa, Carlos Rosell Sr.

For more efficient bioethanol production, the monitoring and control of fermentation processes are crucial for ensuring high ethanol concentrations. However, the nonlinear dynamic relationship between key variables and the inhibitory effects of cellular stress barriers pose challenges in accurately assessing fermentation quality in real time. This study presents a novel approach using machine learning and advanced process analytical technology (PAT) to develop data-driven 'soft sensors' that enable continuous prediction of ethanol and substrate concentrations during intensified ethanol production from sugar cane substrate with cell cycling techniques.

The soft sensor utilizes real-time parameters such as pH, redox potential, capacitance, and temperature to predict ethanol, substrate, and cell concentrations. To overcome data scarcity issues associated with fermentation processes, synthetic time series data generation is integrated into the modeling approach. By employing a variational autoencoder (VAE), synthetic data is successfully generated to enhance the training and testing of deep neural networks on both original and synthetic datasets.

Results from fermentations focusing on intensified ethanol production from sugarcane substrate using cell cycling techniques demonstrate significant improvements in prediction robustness when incorporating generated data. The soft sensor with augmented dataset shows significant increases in robustness and predictive power compared to models trained solely on original datasets. This augmented dataset effectively enhances model generalization by mitigating overfitting issues.

Furthermore, this study highlights that easily measurable PAT tools including pH, redox potential, capacitance, and temperature provide valuable information for inferring key variables reflecting fermentation quality in real time. Feedforward neural networks were optimized for the number of hidden neurons and showed excellent capacity to capture the complex kinetic relationship between the input and the output variables. The developed soft sensor demonstrates promising results in accurately predicting ethanol and substrate concentrations during intensified bioethanol fermentation processes.

Overall, this research showcases an innovative approach that integrates machine learning techniques with advanced PAT tools to enhance the reliability and scalability of bioethanol fermentation monitoring through data-driven soft sensors. The findings pave the way for systematic development towards reliable process optimization strategies for efficient bioethanol production while addressing resource constraints associated with limited real-time measurements

476d Novel Dimethyl Ether Production Technology - Green Alternative for Diesel Engine

Andrzej Rogala

Dimethyl ether (DME) emerges as the simplest yet profoundly versatile ether, heralded for its multifaceted applications as an eco-friendly medium for energy storage and distribution. Traditionally relegated to the role of an aerosol propellant, the breadth of DME's potential utility is vast and varied. Its inherent properties render it an excellent candidate for a myriad of applications, including as an alternative diesel fuel, a precursor in synthesizing a wide range of chemical compounds, including other synthetic fuels, and as a renewable energy source in fuel cells.

The pressing challenges of environmental preservation and the pervasive problem of smog elevate the urgency for cleaner energy solutions. DME's environmental credentials are particularly relevant in this context, offering a path to significantly reduce both particulate matter and gaseous pollutants. Its exceptional physical and chemical characteristics, such as a high cetane number and liquefaction properties akin to LPG, coupled with its combustion process, which remarkably emits no particulate matter, position DME as not merely a potential but a tangible substitute for diesel. This is evidenced by its greater combustion efficiency over diesel, despite DME's almost halved combustion energy. The practical application of DME has been underscored by significant experimental studies in numerous scientific article but also real-world validated in R&D projects.

Further bolstering DME's appeal is its role in marine applications, underscored by its compatibility with stringent environmental regulations, such as the amendment to the Act on the prevention of marine pollution, which incorporates the sulfur emissions reduction directive of the EU. This positions DME as a sulfur-free, environmentally benign fuel choice for maritime transport.

In addition to its immediate applications, DME holds promise as a pioneering chemical hydrogen carrier, facilitating hydrogen production through steam reforming at substantially lower temperatures compared to traditional sources. This aspect of DME not only underscores its versatility but also aligns with global trends towards hydrogen and electric mobility, marking it as a catalyst for the automotive industry's transition from conventional fuel vehicles.

This presentation aims to uncover the extensive potential of Dimethyl Ether (DME) within the international markets, with a focus on the latest advancements in production techniques spanning laboratory and industrial scales. It will offer a comprehensive examination of the forefront in catalyst modification strategies, addressing the significant challenges associated with scaling production processes and navigating the complex pathway to commercialization. A detailed comparative analysis of various reactors used in DME production will be provided, highlighting strategies to enhance production efficiency. Additionally, this study will discuss the role of the proposed technology within the framework of a circular economy, where waste carbon sources are repurposed into valuable fuel. It will also touch upon the necessary shifts in policy and market incentives aimed at fostering the adoption of DME and other green synthetic fuels technologies, presenting a critical dialogue on the integration of sustainable energy solutions in response to global environmental challenges.

The research was financed by the National Center for Research and Development and Polish Oil and Gas Company (PGNiG) under the project 'Development of DME production technology for small hydrocarbon deposits' No. POIR.04.01.01-00-0064/18-00

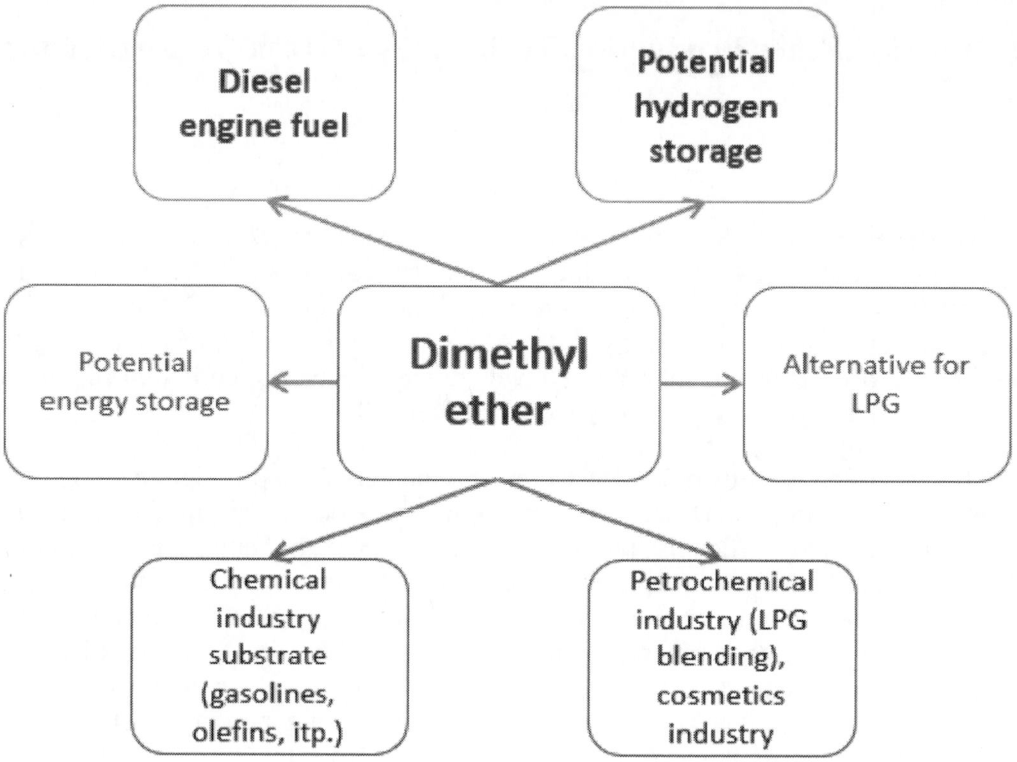
Diesel engine fuel
Potential hydrogen storage
Potential energy storage
Dimethyl ether
Alternative for LPG
Chemical industry substrate (gasolines, olefins, itp.)
Petrochemical industry (LPG blending), cosmetics industry

476e Performance and Cost Analysis of Fueling Heavy-Duty Hydrogen Fuel-Cell Electric Vehicles

Hla Tun, Krishna Reddi, Amgad Elgowainy

The deployment of heavy-duty hydrogen fuel-cell electric vehicles (FCEVs) holds significant potential to enable deep decarbonization through the reduction of fossil fuel consumption and the enhancement of environmental sustainability. For early market penetration of heavy-duty FCEVs, the development of a cost-effective refueling station and its operation strategy is crucial. Unlike light-duty hydrogen FCEV refueling station, heavy-duty hydrogen FCEV refueling station should be able to supply hydrogen at higher mass flow rate which can exceed 60g/s. For H70 station, the peak mass flow rate can be up to 300g/s. So, it is essential to obtain the optimum design of station hydrogen storage systems which can meet the daily hydrogen refueling demand while maintaining the maximum hydrogen utilization and minimum energy consumption. In this study, we use Argonne's Hydrogen Station Cost Optimization and Performance Evaluation (H2SCOPE) thermodynamic model to simulate the operation of heavy-duty hydrogen refueling stations and evaluate the performance of the station operation schemes, including supply storage pressure consolidation and baseline operation with no pressure consolidation. We evaluate the performance of various station configurations and present the performance of the pressure consolidation scheme and typical baseline operation of a station supplied by a high-pressure tube-trailer. We also evaluate the impact of the operation schemes on the refueling cost of hydrogen.

476f Design, Modeling, and Eco-Technoeconomic Analysis of Ammonia Co-Firing in a Natural Gas Combined Cycle Plant

Haoxiang Lai, Lingyan Deng, Guiyan Zang, Angiras Menon, Emre Gencer, Ahmed Ghoniem, Robert J. Stoner

Alternative fuels have drawn attention for their potential to reduce greenhouse gas (GHG) emissions in the power generation sector [1]. Ammonia, as an alternative fuel and a hydrogen carrier, has advantages such as high hydrogen density, ease of long-distance transportation and storage, and compatibility for co-firing with natural gas (NG) [2-4]. However, the feasibility of large-scale application of ammonia co-firing with NG, in terms of techno-economics and life cycle emissions, remains uncertain. In this study, we investigated the viability of both greenfield and retrofit designs of ammonia co-firing in a natural gas combined cycle plant (NGCC) in India through model simulations using Aspen Plus, techno-economic analyses, and life cycle assessments. We explored cases ranging from substituting 20% of NG (in terms of heating value) with ammonia up to 100% pure ammonia combustion, using ten different upstream ammonia production pathways (including grey, blue, and green pathways). For our baseline case, which is greenfield 20% ammonia co-firing in NGCC in India, by using blue ammonia produced via autothermal reforming using air combustion with carbon capture and sequestration in Australia, the levelized cost of electricity (LCOE) and life cycle GHG emission were estimated at \$64/MWh (in 2019 U.S. dollars) and 417 kg-CO_2e/MWh, respectively. Increasing the proportion of ammonia co-firing to 80% resulted in an LCOE of \$121/MWh and a life cycle GHG emission of 324 kg-CO_2e/MWh. The cost of life cycle CO_2 avoided (CCA) for the 20% and 80% ammonia co-firing cases compared to the original NGCC were \$631/tonne-CO_2e and \$617/tonne-CO_2e, respectively. Despite the significant onsite GHG emission reduction achievable with ammonia co-firing (e.g., 80% ammonia co-firing would result in around 80% onsite GHG emission reduction), the CCA remained high due to life cycle GHG emissions or high upstream ammonia costs (\$397/tonne-NH_3 including production and transportation as our baseline case). Considering the anticipated advancements in renewable energy in the future, ammonia co-firing stands to benefit from a reduction in upstream ammonia costs (via green pathways) and its potential as dispatchable power. If the upstream costs of green ammonia (produced via low temperature electrolyzers) reduced to around \$200/tonne-NH_3, the CCA for 20%-80% ammonia co-firing cases can be reduced to around \$100/tonne-CO_2e.

[1] Sánchez, A., Castellano, E., Martín, M., & Vega, P. (2021). Evaluating ammonia as green fuel for power generation: A thermo-chemical perspective. *Applied Energy*, *293*, 116956.

[2] Xiao, H., Howard, M., Valera-Medina, A., Dooley, S., & Bowen, P. J. (2016). Study on reduced chemical mechanisms of ammonia/methane combustion under gas turbine conditions. *Energy & Fuels*, *30*(10), 8701-8710.

[3] Ito, S., Uchida, M., Suda, T., & Fujimori, T. (2020). Development of ammonia gas turbine co-generation technology. *IHI Eng. Rev*, *53*(1), 6.

[4] Ito, S., Uchida, M., Suda, T., & Fujimori, T. (2022). Demonstration Tests of Ammonia/Natural Gas Co-firing Power Generation with a 2-MW-Class Gas Turbine. In *CO2 Free Ammonia as an Energy Carrier: Japan's Insights* (pp. 515-522). Singapore: Springer Nature Singapore.

476h Insights into the Deep-Sea Gas Hydrate Dissociation in the Indian Offshore: An Integrated Study of the Mechanistic, Environmental, and Economic Dynamics

Dibyarup Majumdar, Sanjay M. Mahajani

Natural gas, a relatively cleaner burning fuel, is abundantly stored within natural gas hydrate reserves in the proximate Indian offshore. This attributes these reserves as a potentially sustainable energy prospect for developing nations like India which majorly imports natural gas for its needs. This study aims to develop a model based on first principles to comprehend the deep-water hydrate dissociation dynamics and gauge the gas production rates. Building on the analysis, the study further delves into exploring the environmental and economic ramifications through a detailed Life-cycle assessment (LCA) followed by a comprehensive economic analysis. This undertaking exemplifies a unique attempt to develop such multi-dimensional comprehensions aimed at facilitating the commercialization of the process in the future.

In contrast to most studies in the literature that use commercially available simulators, the modelling methodology in this study uses an equation-based approach that integrates conduction-convection coupled with pressure drawdown for the calculation of dissociation parameters. Furthermore, this study considers the geological and hydrological parameters specific to deep-sea Indian reserves, particularly in the Krishna-Godavari Indian basin. The developed model is simulated by applying the finite-difference technique followed by a sensitivity analysis to evaluate the parametric influence of different hydrological and geological parameters. The results indicate a significant gas recovery potential of approximately 51000 m^3/day over 20 years with a notable sensitivity towards sediment permeation and conductivity.

Conceptualizing a cradle-to-gate scope, the LCA findings for the gas production through hydrate dissociation indicate that the CO_2 emissions is around 1.01 tonnes per tonne of natural gas. Moreover, replacing current natural gas imports (47%) entirely with gas coming in from hydrate dissociation can potentially reduce around 62.5 % of CO_2 emissions. The contribution of such CO_2 emission savings, considering a 100% replacement, can be a significant 4.2% towards the commitment pledged by the Government of India to reduce the country's emissions by 1 billion tonnes by the year 2030. Furthermore, this study discusses several economic aspects of the process through scenarios wherein imported natural gas, procured at spot pricing which is vulnerable to geopolitical fluctuations, is replaced with gas from hydrate reserves, result in savings of about 25 USD/MMBTU.

476i Comparable Techno-Economic Analysis of Wet Waste Hydrothermal Liquefaction with Different Biocrude Upgrading Strategies

Shuyun Li, Yuan Jiang, Daniel Santosa, Karthikeyan Ramasamy, Huamin Wang, Andrew J. Schmidt, Michael R. Thorson

Renewable, low carbon, "drop-in" biofuels offer greater potential to decarbonize aviation, marine, and heavy-duty transportation sectors and play a significant role to meet the zero-greenhouse gas (GHG) emission goal by 2050. Hydrothermal liquefaction (HTL) is one of many promising biofuel technologies, that can convert any wet waste (i.e., sewage sludge, manures, and food waste) to biocrude at high pressure (~20 MPa) and medium temperature (~360°C). It eliminates the need for feedstock drying and produces a high quality biocrude, of which the upgrading/fuel finishing strategies can be optimized and tuned to meet specific needs of the end users for different applications (i.e., marine, aviation and heavy-duty). For marine biofuel production, the upgrading strategy includes mild hydrotreating to remove the heteroatoms, lower total acid number (TAN) for meet fuel standard, as well as using the whole distillates without fractionation as marine fuel. For SAF and renewable diesel production, the HTL biocrude is fully hydrotreated for heteroatoms removal, followed by fractionation into naphtha, jet, heavy diesel/ residue, and hydrocracking of residue and/or heavy diesel. Finally the combined jet fraction from both hydrotreating and hydrocracking went through a deep dehydronitrogenation process to remove the nitrogen below 2 PPM for the SAF requirement. Tuning the distillation column and hydrocracking operating can maximize SAF or renewable diesel production. In this study, techno-economic analyses (TEA) were conducted for a small scale HTL plant (110 dry ton per day waste) with a centralized upgrading plant (about 3000-barrel biocrude per day) based on rigorous process model developed in Aspen Plus and experimental data collected from lab-scale continues flow system. A variety of biocrude upgrading configurations were explored for maximizing marine, SAF and renewable diesel production, respectively.

206a Plant-Wide Modeling and Techno-Economic Analysis of Microwave-Assisted Gasification Process for H_2 Generation Using a Mixed Feedstock of Biomass and Plastics

Md Emdadul Haque, Opeyemi Ogunniyan, Pranjali Muley, Ashraf Abedin, Dushyant Shekhawat, Debangsu Bhattacharyya

Projected future consumption of H_2 is anticipated to reach 2.3 Gt/yr across industries including synthetic chemistry, transportation, heating, and power generation [1]. Hydrogen holds the potential to decarbonize approximately 18% of energy-related sectors and playing a crucial role in transitioning towards a fully renewable energy society [2]. Currently, over 98% of H_2 is derived from fossil fuels, predominantly through processes like steam reforming (76%), and coal gasification (22%) [3]. To mitigate rising energy demands, environmental concerns, depletion of resources, and volatile fossil fuel prices have spurred interest in producing H_2 from sustainable energy sources such as biomass, waste plastics, and municipal solid waste. To generate H_2 from solid feedstock (biomass and mixed plastics), gasification, a thermochemical conversion process, is a preferred method. However, gasification process is energy intensive, requires high temperature up to 1100°C, making the process infeasible. In recent times, there has been growing interest in microwave assisted heating process within chemical reaction systems due to their low energy requirements compared to conventional methods [4]. As microwave directly interacts with the material, temperature increases rapidly due to selective heating, enhancing overall energy efficiency and reducing operational costs. Furthermore, nonthermal effect of microwave helps to limit the side reactions and can improve the yield of desired products. This study focuses on developing a plant-wide model for producing H_2 from a mixed feedstock of biomass and mixed plastics. Initially, a lab-scale air-assisted microwave gasifier model is developed by applying data reconciliation technique to minimize errors with experimental results. This lab-scale model is then scaled up to a commercial level and integrated into the plant-wide model. The plant wide model consists of several sections including gasification, gas cleaning, cooling, scrubbing, compression, sulfur removal, steam reforming, shift conversion, and H_2 purification using pressure swing adsorption (PSA) [5]. Economic analysis of the plant-wide model shows a reduction in capital expenditure (CAPEX) and operating expenditure (OPEX) by 10% - 15%, specifically 35% reduction in energy requirements and a 20% improvement in net present value (NPV) compared to conventional process. Sensitivity studies are performed for variation in feed mix and its composition.

References:

1. Rosyadi, I.; Suyitno, S.; Ilyas, A.X.; Faishal, A.; Budiono, A.; Yusuf, M. Producing hydrogen-rich syngas via microwave heating and co-gasification: a systematic review. *Biofuel Research Journal*, **2022**, 33, 1573-1591.
2. Oliveira, A.M.; Beswick, R.R.; Yan, Y. A green hydrogen economy for a renewable energy society. *Opin. Chem. Eng.*, **2021**, 33, 100701.
3. Lepage, T.; Kammoun, M.; Schmetz, Q.; Richel, A. Biomass-to-hydrogen: a review of main routes production, processes evaluation and techno-economical assessment. *Biomass Bioenergy*, **2021**, 144, 105920.
4. Abedin, A.; Bai, X.; Smith, M.; Muley, P. Microwave-assisted co-gasification of mixed plastics and corn stover: A synergistic approach to produce clean hydrogen. *Energy Conv. Man.*, **2023**, 280,116774.
5. Spath, P.; Aden, A.; Eggeman, T.; Ringer, M.; Wallace, B.; Jechura, J. Biomass to Hydrogen

Production Detailed Design and Economics Utilizing the Battelle Columbus Laboratory Indirectly-Heated Gasifier. *National Renewable Energy Laboratory*, **2005**, CO, USA.

206b Closing the Loop on Polystyrene Using Chemical Recycling: Thermodynamic, Economic, and Emissions Analysis

Michael Timko, Madison R. Reed, Bernardo Castro Dominguez, Elizabeth Belden, Nikolaos Kazantzis

Polystyrene is used in food packaging and shipping where its mechanical and thermal properties are difficult to replicate in cost effective competitive materials. Unfortunately, less than 5% of polystyrene is currently recycled due to the many challenges associated with collecting, sorting, and especially processing the end-of-life material. Interestingly and unlike other major polymers, polystyrene unzips during pyrolysis to produce roughly 60% yield of the styrene monomer. This monomer is contaminated with other 1-ring side products, including toluene, ethyl benzene, and a-methyl styrene. Removing these impurities provides an opportunity to recover monomer-grade styrene that can be repolymerized to close the loop on polystyrene use. In this study, we analyzed a reaction-separation process to determine the thermodynamic, economic, and emissions performance. Figure 1 is a schematic. Thermodynamically, the process can produce monomer-grade styrene (>99% purity) with less than 10 MJ/kg energy input for many relevant scenarios, as shown in Figure 2. Economically, we find that realistic prices of waste polystyrene, ranging from -60 to +60 USD per ton result in positive investment opportunities, as measured by the net present value (NPV) and as summarized in Figure 3. Combining the thermodynamic analysis with the economic analysis permits estimation of a CO2 abatement cost. Using capital expenses as the resource input and comparing with pyrolysis to fuels as competing approach to avoid landfilling results in an abatement cost of about \$1.50 per ton of averted CO_2.

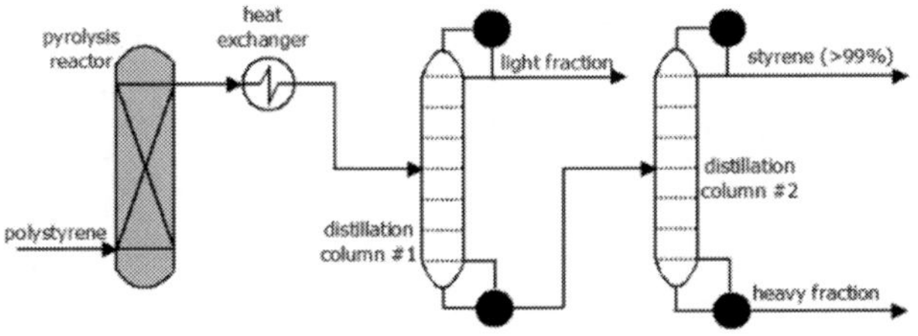

Figure 1. Schematic of the reaction-separation process that was analyzed in this study.

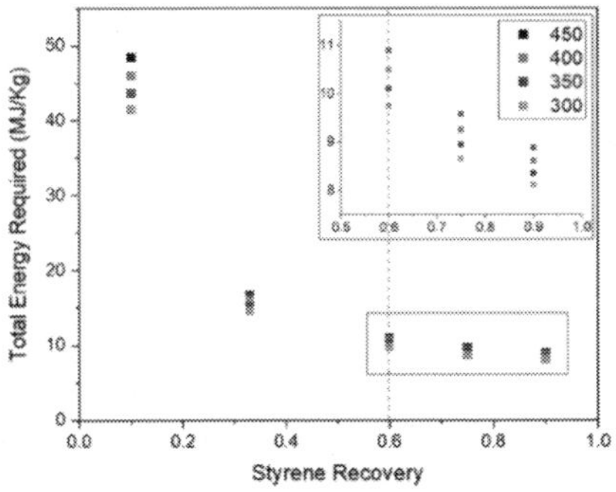

Figure 2. Energy required for recovering >99% pure styrene for different pyrolysis temperatures and for different styrene yields. The vertical line marks 60% yield, which has been reported in the literature.

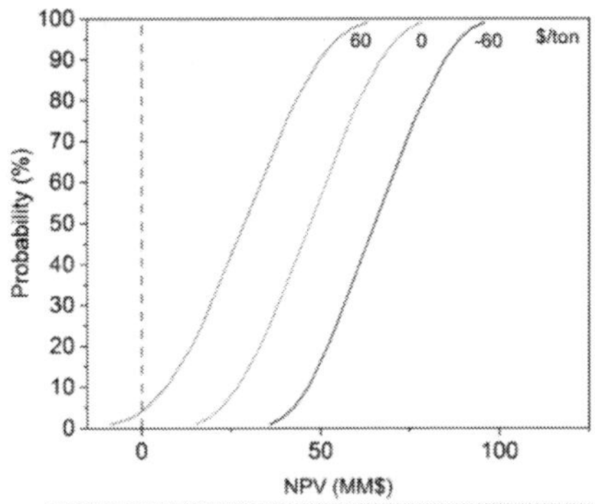

Figure 3. Net present value obtained from Monte Carlo analysis of several different realistic feedstock cost values (-60, 0, and +60 \$/ton). The vertical line indicates a break-even proposition.

206c Plastic Derived Oil and Bio Feedstock Integration into Refineries: The Key to Unlocking Circular Polymers at Scale

Rasha Daadoush, Maria Soliman, Vishvedeep Bhat, Nicolas Goyheneix

Society is currently facing the dual challenge of minimizing plastic waste while also aiming to reduce global greenhouse gas emissions. Aramco and Sabic's shared vision for the future is that plastic should not end up in the environment, landfills or in our oceans and instead is reused and remade into new products.

We are driving the efforts to create a circular global society, by working with consumers, retailers, recyclers and manufacturers to enable valuable materials to be collected and processed to make new products and identity key large scale plastic sinks while demonstrating that closing the loop on used plastic is achievable. The aim is to produce one million tons of circular products per year by the end of the decade.

We will highlight our key efforts to advance the chemical recycling of hard to recycle mixed plastic waste and introduction of renewable bio based feedstocks to produce lower carbon certified circular polymers. These efforts include the development of novel decontamination, depolymerization and conversion technologies that convert mixed plastic waste into feedstock for refinery catalytic and non catalytic conversion reactors and ultimately for conversion into high value circular plastics through introduction into steam crackers.

We will also highlight our efforts to address the high concentrations of contaminants (chlorine, metals, oxygenates and aromatics) that lead to corrosion and potential catalyst poisoning and the novel approaches we are considering to address these contaminants for large scale integration of plastic waste into refineries and chemical plants.

206d Sustainable Aviation Fuel from Municipal Solid Waste Via Methanol Pathway

Francesca Pierobon, Steven Phillips, Dilara Goreke, Udishnu Sanyal, Mond F. Guo, Karthikeyan Ramasamy, Yuan Jiang

The U.S. aviation sector has adopted a long-term aspirational goal of net-zero carbon emissions by 2050. Due to aircrafts weight and size constraints, electrification may not be a viable option for commercial aviation in the mid-term. Therefore, the sector will heavily rely on the use of Sustainable Aviation Fuel (SAF) to meet its emission reduction goal with an expected market of 3 billion gallons per year by 2030 and 35 billion gallons per year by 2050. The growing demand for SAF will require a diversification in SAF production by identifying innovative pathways to leverage alternative feedstocks.

Producing SAF from oxygenated intermediates, especially renewable methanol, has gain increasing interest in the past few years. Unlike other alcohols, renewable methanol can be efficiently produced from a variety of alternative renewable carbon sources using syngas as the intermediate.

In this study we analyze the feasibility of producing SAF from municipal solid waste (MSW). The pathway includes conversion of MSW to syngas via gasification, production of methanol from syngas, conversion of methanol to olefins, subsequent oligomerization of olefins, olefin hydroprocessing for fuel production, and separation of final products. We propose an innovative catalytic approach characterized by a bifunctional Ni/SiO_2-Al_2O_3 (metal)/Zeolites (acid) catalyst for single-step co-oligomerization of C_2 and C_{3+} olefins. The proposed approach has a demonstrated single-pass conversion of approximately 92% for C_2 and over 97% for $\geq C_3$ olefins, and a selectivity of about 74% towards the production of jet-range products. A chemical process simulation, techno-economic analysis and life cycle assessment are performed to evaluate the technical, economic, and environmental feasibility of the proposed process.

206e Life Cycle Analysis of Closed Looped Plastics Waste to Olefin Via Process Simulation

Wajid Shaik, Ghoncheh Rasouli

Problem / Scope

The global production of plastics exceeds 380 million tons annually, with only 16% being recycled into feedstock. Waste plastics significantly contributes to environmental pollution and climate change through emissions with 40% of waste plastic sent to landfills, 25% to incineration, and 19% being dumped. Effective plastic waste management is crucial to mitigate these issues, necessitating strategies that emphasize waste reduction, recycling initiatives, and innovative technologies to minimize environmental impact and promote sustainability.

The escalating volume of plastic waste and the environmental concerns associated with traditional disposal methods have created a strong market demand for innovative solutions like plastic pyrolysis. This demand is further fueled by increasing consumer awareness and government initiatives to promote recycling and reduce plastic pollution.

Continuous efforts from companies and research institutions are directed towards the development and refinement of innovative plastic pyrolysis technologies. This involves a comprehensive understanding of complex kinetics, reactor design, and the impacts of optimum operating conditions on desirable key performance indicators.

A key component of plastic waste management is chemical plastic recycling, which helps emission reduction to net zero and circularity of the plastics value chain through recycling and processing end products to bio- fuels. Thermal and catalytic pyrolysis of plastic waste have emerged as promising methods, not only for environmental benefits but also for reducing the consumption of fossil fuel resources.

Methods

In the process of producing olefin from plastic waste, medium-temperature thermal pyrolysis yields liquid hydrocarbon feedstock for steam cracking, while high-temperature pyrolysis generates light gases suitable for olefin production. Subsequently, plastic oil hydro-processing is utilized to remove impurities, thereby ensuring the integrity of downstream processes such as steam cracking. The non-catalytic pyrolysis of plastic waste involves thermal cracking of long heavy-chain hydrocarbon molecule into smaller, more valuable ones.

Despite the ongoing development, the commercialization of plastic pyrolysis technologies has yet to mature. Process simulation serves as a valuable tool for stakeholders, including technologists and engineers, providing insights and analysis into the circularity of olefins and polymers.

Results and Implications

A process simulation model was developed using KBC's Petro-SIM Simulation modelling software to analyze the closed-loop olefin production to polymer products, then conversion of plastics waste to olefin.

This study investigates the modeling and simulation of the entire circular production chain, encompassing the transformation of plastic waste into their building block and other valuable hydrocarbons. The study explores the energy efficiency and emission reduction potential within the circular economy framework of plastic management. By converting plastic waste into synthetic resources, sustainable plastics can be produced for various industries, including food packaging, pharmaceuticals, and textiles. This work contributes to the advancement of sustainable practices and enhances the resilience of the plastics industry amid environmental challenges.

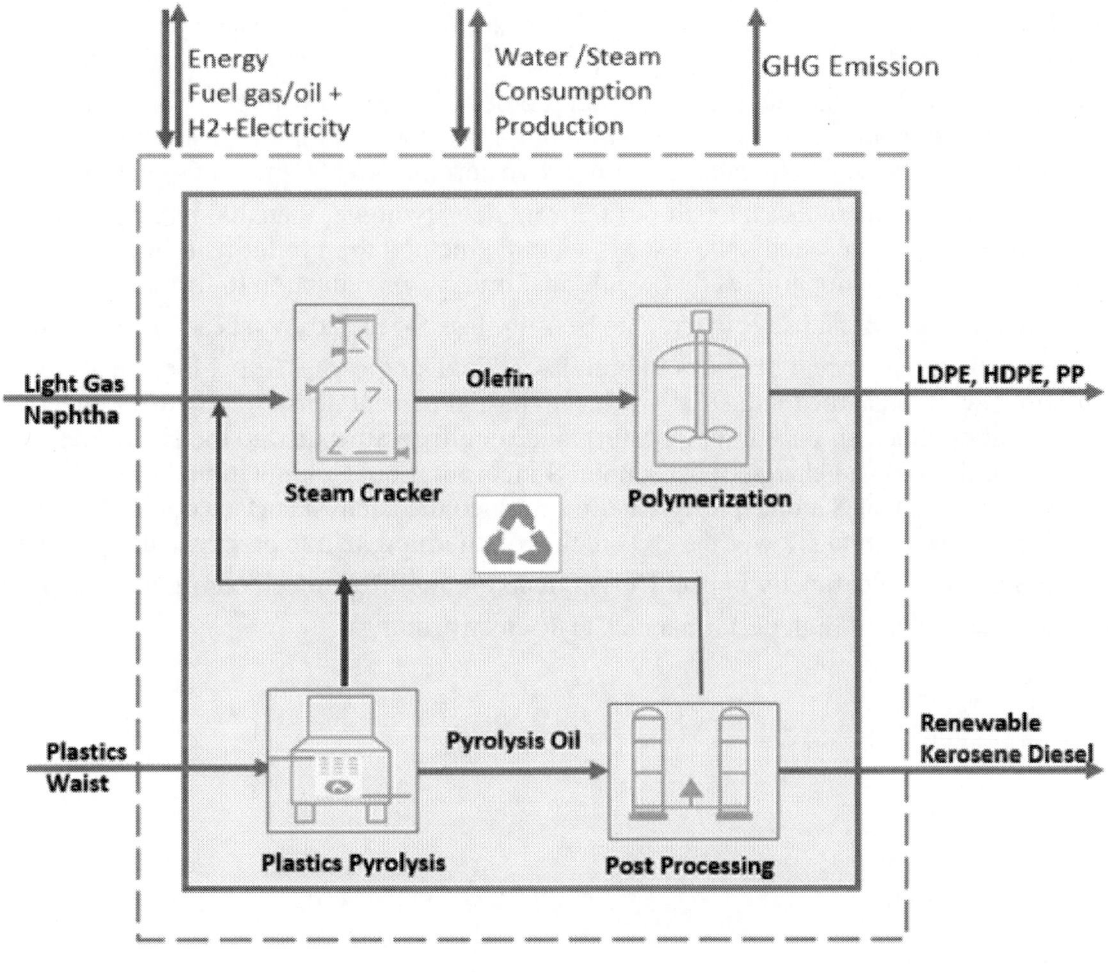

206f How Much Is Surface Dopant Enough to Maximize CO_2-to-Liquid Chemicals Conversion at Industrially Relevant Current Density?

Thuy Duong Nguyen Phan, James Ellis, Anantha Venkataraman Nagarajan, Bret H. Howard, Giannis Mpourmpakis, Douglas Kauffman

Electrochemical reduction of CO_2 (CO_2R) to valuable, carbon-neutral chemical feedstocks and storable fuels driven by renewable electricity has been recognized as one of the promising technologies to mitigate the greenhouse effect, reduce global demand for fossil fuels, create sustainable energy, and achieve net zero emission by 2050. Liquid formic acid or formate electrochemically produced via CO_2R has been identified as an economically sustainable, green hydrogen carrier and fuel that can reduce carbon footprint compared with gray fossil-based formic acid. Tremendous ongoing efforts have focused on tuning catalyst morphology, composition, structure, size of tin-based electrocatalysts to improve current density, product selectivity, and durability to approach the industrial viability (current density higher than 200 mA/cm^2 and lifetime beyond 1,000 hours). In this study, we discuss a simple method to introduce surface sulfur heteroatoms into SnO_2 nanoparticles that would maximize CO_2 reduction to formate/formic acid with impressively high current and selectivity. The best-in-class S-SnO_2 catalysts achieved some of the highest reported partial current densities in both H-cell and single-gap, full-cell electrolyzer configurations while sustained industrially relevant partial current density and ~80% selectivity over hours of operation. Such an outstanding performance could be attributed to the electronic structural modification and improved charge transfer upon S incorporation. Computational modeling additionally revealed that S atoms preferentially occupied the catalyst surface could increase key intermediate stabilization and lower the CO_2 reduction thermodynamic barrier. Our demonstration of how dilute S dopants beneficially impact CO_2 conversion in different cell configurations provides further design concepts of high performance CO_2R electrocatalysts.

206g Catalytic Hydrodeoxygenation of Bio-Oil Produced from Co-Pyrolysis of Biomass and PET Plastic Waste

Janaki Komandur, Kaustubha Mohanty, Vinu Ravikrishnan

The present study investigates the catalytic hydrodeoxygenation of bio-oil produced from co-pyrolysis of Mesua ferrea L (commonly known as Nahar, a non-edible oilseed) and PET plastic. The operating parameters, such as temperature and biomass: plastic feedstock ratio, were optimized, and it was found that maximum bio-yield (35 wt%) was obtained at 873 K, 2:1 (biomass: plastic). The qualitative analysis of the co-pyrolyzed oil revealed that the fatty and carboxylic acids (~55%) dominated the composition. Hence, the bio-oil was upgraded for future applications. In the upgradation process, red mud (RM), an alkaline waste generated during the Bayer process of alumina production, was utilized as a support for mono (Ni) and bi-metallic catalysts (Ni-Co, Ni-Mo) at various concentrations of nickel (5 wt.% each). The catalyst was characterized using FESEM-EDX, FETEM, XRD, ICP-MS, XPS, and Surface area analysis. These catalysts were employed in the hydrodeoxygenation (HDO) of co-pyrolytic oil. The activity of bi-metallic catalysts for HDO reactions increased the PIONA (paraffin, isoparaffins, olefins, naphthenes, aromatics). The highest organic liquid yield (72%) was achieved using the Ni-Co/RM catalyst with low coke formation and high hydrocarbon yield (80%). The resultant upgraded oil exhibited an oxygen content of less than 2wt.%. The properties of the HDO oil were obtained using mono and bi-metallic catalysts at a reaction temperature of 513 K, which ultimately reduced the raw bio-oil's acid content (~55%) and converted to hydrocarbons (80%). The primary reaction mechanism that followed was hydrodeoxygenation (HDO) and decarboxylation (DCO). Overall, applying Ni (10wt%), Ni-Co, and Ni-Mo supported Red-mud catalysts improved HDO reactions.

206h Production of Propylene from Polypropylene Waste Via Direct Microwave Catalytic Upcycling

Chunlin Luo, Brandon Robinson, Jianli Hu, Yuxin Wang

Plastic waste crisis has led to severe environmental pollution and an obvious imbalance between production and upcycling. Although conventional catalytic depolymerization can be used for converting plastic waste into chemicals, it remains a significant challenge to achieve high selectivity towards value-added monomers using a one-step method under mild conditions. In this work, we report the high yield of propylene from polypropylene waste through a direct microwave-assisted catalytic process under mild conditions (ambient pressure, 400°C). The conversion of polypropylene reaches 100% with propylene yield up to 55 wt.% using Ru/Fe_2O_3 as a catalyst at 400°C under microwave irradiation. In contrast, there is no conversion of polypropylene under thermal conditions at 400°C. As the reaction temperature under thermal conditions is increased to 600 °C, polypropylene conversion reached 100%, albeit with a lower propylene yield of 43 wt.%.

In addition, the reaction mechanism of upcycling polypropylene waste into propylene under microwave irradiation was analyzed by in-situ FTIR, in-situ Raman, and DFT calculation, as shown in **Figure 1**. Meanwhile, the microwave heating behaviors were studied by COMSOL modeling.

The combination of microwave-assisted catalytic upcycling and efficient microwave-sensitive catalyst is a promising approach to revolutionize plastic waste upcycling, turning a global environmental challenge into an industrially and economically valuable resource.

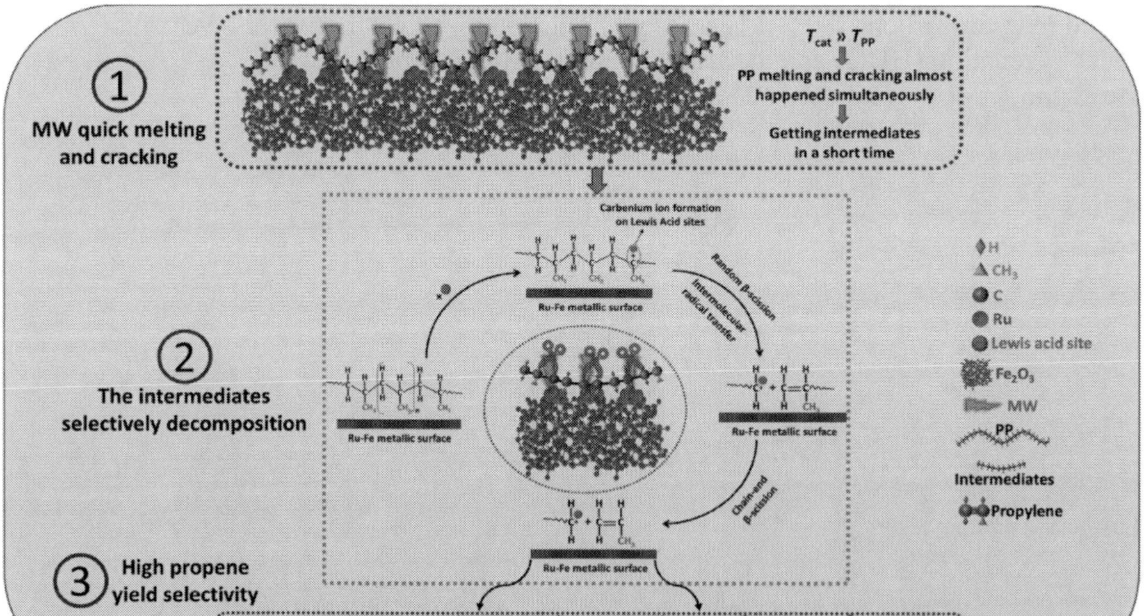

645a Modeling and Simulation for the Natural Gas Midstream Value Chain

Qiang Xu, Parthkumar Babariya, Jatin Sutariya

The U.S. oil and gas midstream market will continue increasing significantly in the coming years, bringing tremendous business opportunities to the midstream industry. However, the industry is also facing enormous environmental challenges and renewable energy competitions in the current era of industrial decarbonization. To study the healthy and sustainable development of the industry, this paper has developed a digital twin a general natural-gas midstream value chain (NGMVC). It includes rigorous process models from raw gas compression, gas/liquid separation, mercury removal, sweetening, dehydration, NGL recovery, and NGL fraction to LNG liquefaction and helium recovery. Other critical industrial sectors such as sulfur recovery and LNG storage are also modeled. Based on the NGMVC model, potential opportunities for enhancing sustainability the NGMVC are ready to be investigated. The study will provide scientific knowledge, data, and valuable technical support for the long-term sustainable development of the natural gas midstream industry.

645d Technoeconomic Framework of Natural Gas Pyrolysis As an Alternative to Flaring and Venting in the Oil Extraction Sector

Silvia Pelucchi, Paolo Mocellin, Federico Galli

Among the world's leading oil and gas producers there is Canada. Canadian extraction sites are dispersed in a broad region, almost as big as Europe's surface (< 140000 km^2), corresponding to Alberta and Saskatchewan provinces [1].

Every time oil is extracted, generally through fracking technology, the light hydrocarbons present on top of the bituminous mass in the form of natural gas are usually flared or, even worse, vented. This is done because of the scattered nature of the extractive sites and because of their location, miles away from the main inhabited areas. Building pipelines and infrastructures to collect the natural gas - which is less than 5% of the total bitumen production - is unfeasible, as it would be very expensive, considering the required length and the necessity of having pressurized pipelines [2].

Flaring and venting practices represent a severe issue for Canada, both economic and environmental. In fact, 1.7% of the overall gas emissions are due to venting [3].

Natural gas is mostly composed of methane, which is the second highest contributor to the greenhouse effect. Its global warming potential is 37 times that of CO_2 [4]. To partially solve the situation, a stricter taxation system, the so-called carbon tax, has lately been introduced, which restricts total carbon emissions and obliges oil producer to reduce their emission by 1% per year [5]. Parallel to this, some new technologies have been developed to either trap or convert these gases, mainly into hydrogen. The gas-to-liquid (GtL) technologies are among the most common, along with steam methane reforming (SMR) and methane pyrolysis.

In particular, SMR provides approximately 48% of the global H_2 demand [6]. SMR allows the production of the so-called grey H2. One main issue connected with this H_2 production is that CO_2 is also produced [7].

An alternative solution is outfitting the plant with Carbon Capture and Storage (CCS) technologies to mitigate the emissions, but this often represents a prohibitive additional cost [6], [8].

In this context, decarbonizing fossil fuels by recovering and sequestering solid carbon instead of gaseous CO_2 can be a suitable alternative [9], [10].

Methane pyrolysis splits CH_4 directly into its components, i.e., hydrogen and carbon. Unlike other technologies involving fossil fuels, the major benefit of methane pyrolysis is the production of CO_2-free hydrogen. Theoretically, this is the best possible option since it is not as endothermic as steam reforming and produces two valuable commodities: the so-called *turquoise* hydrogen and solid carbon. Notably, the latter is not emitted and therefore not transformed into CO that becomes CO_2 afterward. This route is particularly attractive, but there are some technical challenges related to commercialization. Particularly, the low cost of turquoise hydrogen and the uncertainty around C as a source of income are still research questions.

Here we aim to demonstrate that methane pyrolysis is a viable and potentially profitable technology in the context of Canadian remote extraction sites.

Here we propose the technical and economic framework of an innovative process, performed *in situ* to avoid flaring and venting of natural gas. The proposed process is based on non-catalytic methane pyrolysis at T = 950 °C and atmospheric pressure, performed directly at the extraction site, converting the gas into solid carbon and hydrogen.

Part of produced hydrogen is used as a fuel to sustain the process energetically, while solid carbon can find many applications in industry as a reducing agent, in the production of batteries, or in soil amendment and environmental remediation. Two configurations were conceptualized and simulated in Aspen Plus. In Configuration 1, 39 % of H_2 is burnt to sustain the reactor, while the main part is compressed and sent to a turbine to produce electricity. In Configuration 2, 23 % we admit the possibility of selling the hydrogen that is not used to sustain the reactor. In this case, a PSA is introduced to obtain 99.9 % pure hydrogen.

For each Configuration economic indicators were calculated, and environmental issues were discussed to identify the most promising configuration given the current context. Both configurations reduce CO_2 emissions by more than 92 % compared to flaring.

From a profitability perspective over 10 years, Configuration 1 is characterized by a Simple Payback Period (SPBP) of 2.6 years and an IRR of 25.2 %, while they are equal to 3.2 years and 19.5 % for Configuration 2.

References

[1] A. K. Katta *et al.*, "Assessment of energy demand-based greenhouse gas mitigation options for Canada's oil sands," *J Clean Prod*, vol. 241, p. 118306, 2019, doi: 10.1016/j.jclepro.2019.118306.

[2] Alberta Energy Regulator, "Upstream Petroleum Industry Emissions Report - Industry Performance for Year Ending December 31, 2020," 2021. [Online]. Available: www.aer.ca

[3] Environment and Climate Change Canada, "NATIONAL INVENTORY REPORT 1990-2020: GREENHOUSE GAS SOURCES AND SINKS IN CANADA," 2022.

[4] F. Galli, J.-J. Lai, J. De Tommaso, G. Pauletto, and G. S.Patience, "Gas to Liquids Techno-Economics of Associated Natural Gas, Bio Gas, and Landfill Gas," *Processes*, pp. 1–15, 2021.

[5] Government of Canada, "How carbon pricing works." Accessed: Mar. 20, 2024. [Online]. Available: https://www.canada.ca/en/environment-climate-change/services/climate-change/pricing-pollution-how-it-will-work/putting-price-on-carbon-pollution.html

[6] N. Sánchez-Bastardo, R. Schlögl, and H. Ruland, "Methane Pyrolysis for CO2-Free H2 Production: A Green Process to Overcome Renewable Energies Unsteadiness," *Chem Ing Tech*, vol. 92, no. 10, pp. 1596–1609, 2020, doi: 10.1002/cite.202000029.

[7] N. Muradov, "Low to near-zero CO2 production of hydrogen from fossil fuels: Status and perspectives," *Int J Hydrogen Energy*, vol. 42, no. 20, pp. 14058–14088, 2017, doi: 10.1016/j.ijhydene.2017.04.101.

[8] S. Pelucchi, F. Galli, C. Vianello, and P. Mocellin, "Integrating the Benefits of Turquoise Hydrogen to Decarbonise High-Emission Industry," *Chem Eng Trans*, vol. 105, pp. 25–30, 2023, doi: 10.3303/CET23105005.

[9] T. Li, J. Long, W. Du, F. Qian, and V. Mahalec, "Three pathways towards elimination of CO2 emissions from industrial plants that use hydrocarbon fuels," *J Clean Prod*, vol. 391, p. 136159, Mar. 2023, doi: 10.1016/j.jclepro.2023.136159.

[10] S. M. Ali, I. I. I. Alkhatib, A. AlHajaj, and L. F. Vega, "How sustainable and profitable are large-scale hydrogen production plants from CH4 and H2S?," *J Clean Prod*, vol. 428, p. 139475, Nov. 2023, doi: 10.1016/j.jclepro.2023.139475.

645f Magnetic Surfactants for Cost-Effective Control of Shale Swelling in Drilling Fluids

Syed Muhammad Shakil Hussain, Muhammad Shahzad Kamal, Mobeen Murtaza, Shirish Patil

During drilling operations, shale swelling poses significant challenges due to its interaction with water, leading to a range of issues that are often difficult to manage. Shale swelling inhibitors are routinely added to drilling fluids to mitigate these problems. Among these additives, surfactants are widely employed to prevent shale swelling, reduce bit balling issues, and enhance lubricity. In this groundbreaking study, we introduce a novel magnetic surfactant engineered specifically for inhibiting shale swelling in drilling fluids. The evaluation of shale swelling inhibition involved a comprehensive set of tests, including linear swelling tests, capillary suction timer tests, zeta potential measurements, and scanning electron microscopy. The performance of the newly developed magnetic surfactant was compared with traditional shale stabilizers, such as potassium chloride (KCl). Two distinct concentrations of magnetic surfactants (0.2% and 0.4% by weight) were employed to prepare the investigative solutions. Additionally, rheological and fluid loss tests were conducted to assess the impact of magnetic surfactants on drilling fluid properties. Remarkably, the results demonstrated a significant reduction in shale swelling when magnetic surfactants were introduced. Moreover, the addition of magnetic surfactants to clay led to noticeable improvements in fluid loss and rheological properties, indicating reduced clay hydration. The zeta potential shifted from highly negative to a lower negative value, and the capillary suction timer readings decreased. These observations underscore the effectiveness of magnetic surfactants in providing long-term shale swelling inhibition by intercalating into clay layers and preventing their interaction with water. This study marks a pioneering effort in the application of magnetic surfactants as inhibitors of clay swelling in drilling fluids. The experimental findings highlight magnetic surfactants as a promising and cost-effective alternative to conventional swelling inhibitors, offering innovative solutions to mitigate shale-related challenges in drilling operations.

535a Large-Scale CO_2 Utilization for Production of Methanol in Carbon-Intensive Jurisdictions

Yurley Karina Anaya Jaimes, Jubil Joy, Amit Kumar

This study explores the potential of large-scale CO_2 utilization to produce methanol in regions heavily reliant on carbon-intensive energy sources, such Alberta, a western province in Canada. There is an urgent need to mitigate carbon emissions and transition to cleaner energy alternatives. Thus, the main objective is to assess the techno-economic feasibility and environmental impact of using CO_2 hydrogenation processes for methanol synthesis, with a focus on jurisdictions with substantial carbon footprints. The scope extends to evaluating the efficiency improvements, cost implications, and potential carbon reduction.

The methodology includes comprehensive process modeling. The models are developed to represent the methanol synthesis process through CO_2 hydrogenation reactions, including different scenarios and operational parameters. CO_2 is sourced from atmospheric air using direct air capture (DAC), a carbon removal technology. The study focuses on the large-scale production of methanol, 7,000 tonnes per day, via CO_2 hydrogenation, and its comparison to methanol production via conventional natural gas-based technologies, steam methane reforming (SMR) and autothermal reforming (ATR). CO_2 hydrogenation for methanol production includes DAC unit and hydrogen production. This research provides a detailed understanding of the methanol production process, identifies optimal operating conditions, and estimates the potential for carbon emission reduction. A sensitivity analysis is also included to understand the impact of variable feedstock prices, energy costs, and carbon pricing mechanisms on the overall feasibility of the process. The results provide a clear picture of the potential efficiency gains, cost benefits, and environmental impacts associated with hydrogen utilization in methanol production within carbon-intensive jurisdictions.

The study makes a novel contribution by providing region-specific insights and recommendations that would benefit jurisdictions with high carbon footprints. The comprehensive modeling approach and utilization of simulation model offers a robust foundation for evaluating the large-scale production of methanol. It ensures a thorough and accurate analysis, setting the basis for informed decision-making and sustainable industrial practices. The study focuses on establishing a clear link between CO_2 utilization, economic viability, and carbon emission reduction, for the adoption of CO_2 -based technologies in carbon-intensive regions.

535b Highly Selective Synthesis of Liquid Petroleum Gas from CO_2 Hydrogenation over Czza/β-Zeolite Catalyst

Kaiying Wang, Shiguang Li, Miao Yu, Xinhua Liang

The selective hydrogenation of CO_2 to produce valuable liquid petroleum gas (LPG) is a desirable yet challenging process due to the concurrent occurrence of numerous competing reactions. Herein, we report the development of a $Cu/ZrO_2/ZnO/Al_2O_3$ (CZZA) and β-zeolite composite catalyst that exhibited remarkable LPG selectivity of 80.3% among hydrocarbons (not including CO) at a CO_2 conversion rate of 34.5% under relatively mild conditions of 300°C and 2 MPa in a fixed bed reactor. Notably, a significant LPG yield of 19.5% was achieved, outperforming the majority of previously reported studies. Furthermore, this exceptional catalytic performance was maintained for an extended period of 160 hours, demonstrating excellent stability. Remarkably, an even higher LPG yield of 23% was obtained by increasing the pressure to 3 MPa, owing to the promotion of secondary hydrogenation reactions over the β-zeolite and the inhibition of CO formation under a higher pressure. This research provides a promising strategy for the highly selective synthesis of LPG from CO_2 hydrogenation, offering a potential solution to convert greenhouse gases into valuable liquid fuels.

DEVELOPMENT OF MODELS FOR THE ESTIMATION OF NET HEAT OF COMBUSTION OF AVIATION FUELS

Hasan Mashni, German Aerospace Center, Stuttgart, BW, Germany
Benedict Enderle, German Aerospace Center, Stuttgart, BW, Germany
Uwe Bauder, German Aerospace Center, Stuttgart, BW, Germany
Georg Eckel, German Aerospace Center, Stuttgart, BW, Germany
Patrick Le Clercq, German Aerospace Center, Stuttgart, BW, Germany

Hasan Mashni is a Diploma Engineer with expertise in applying advanced technologies to address real-world challenges related to sustainability and climate change. He is currently based at the German Aerospace Center (DLR) - Institute of Combustion Technology. His background spans research experience at the Center for Applied Geosciences, University of Tübingen, as well as various companies and start-ups. Hasan's research interests encompass machine learning, mobile sensor systems, and numerical modeling. Recent projects have focused on sustainable aviation fuels, atmospheric physics, turbulence modeling, and computational fluid dynamics.

INTRODUCTION

The rapid development of sustainable aviation fuels (SAFs) and the constantly expanding landscape of approved or in-research aviation fuel production pathways, as well as their blends, come with several operational and safety challenges that limit the potential environmental and economic benefits of such technologies. Many of these challenges fall within the scope of quantifying the composition and properties of these fuels to ensure their compliance with specifications following ASTM D1655 [1] or D7566 [2]. While standard methods for measuring these quantities are well established, the associated cost, operational overhead, and required expertise limit their application to laboratory conditions e.g., net heat of combustion - ASTM D4809 [3], freezing point - ASTM D2386 [4], aromatic content - ASTM D1319 [5] & D5769 [6], etc., making them difficult to utilize in real-time field applications. Although machine learning (ML) models based on the detailed composition of fuels can help reduce a significant amount of effort and complexity associated with measurement (e.g., for prescreening SAF candidates [7]), similar limitations restrict the field application of these models due to their dependence on comprehensive gas chromatography techniques [8]. Therefore, to address this challenge, empirical or chemometric models based on property correlations are studied, which have the potential to simplify and improve the analysis of complex aviation fuels with various compositions.

This study focuses on comparing, developing, and simplifying property-based empirical or chemometric models for estimating the properties of both conventional and sustainable aviation fuels, specifically focusing on models used to estimate the net heat of combustion (NHOC). Such models are developed based on long-standing chemical expertise and a large dataset of precise fuel property measurements. They are primarily valid for well-defined conventional jet fuels based on crude oil that are well represented in the training dataset of such models. Therefore, the generalization of these models to alternative fuels is bounded by two challenges. Firstly, considering the statistical nature of these models and new variations in the composition of SAF, such models could become outdated and, unless retrained to accommodate the new variations, lose accuracy [9][10][11]. Secondly, these models require the precise measurement of difficult-to-measure properties such as distillation curves or aromatic content, which make their field application difficult (example models: net heat of combustion - ASTM D3338 [12], flash point [13], hydrogen content - ASTM D3343 [14], calculated cetane index - ASTM D976 [15] and D4737 [16], etc.).

METHODS

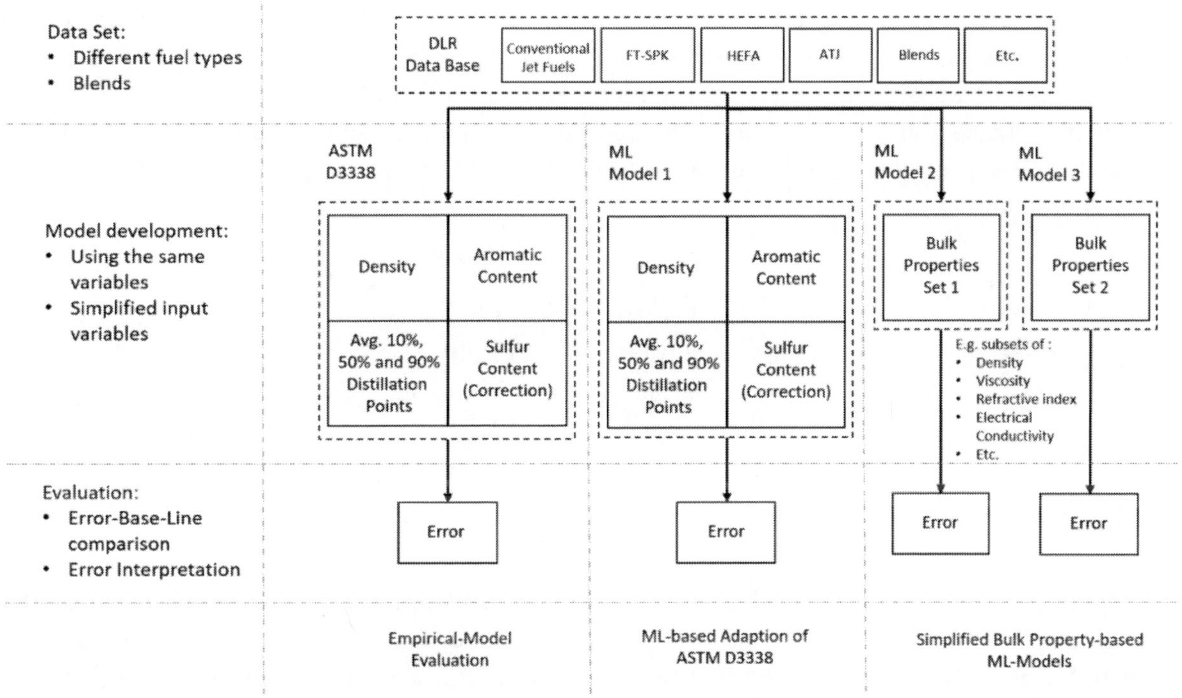

Figure 1: Comparison scheme of the net heat of combustion estimation methods

In this context, this study focuses on property-based models for NHOC estimation that utilize NHOC correlations with different aviation fuel properties. NHOC is chosen for its potential to drive both environmental and economic incentives, thereby highlighting the urgency of sustainable aviation fuel (SAF) development. Furthermore, because NHOC correlates with both bulk properties and fuel composition, it serves as an excellent basis to demonstrate the approaches developed in this study across different fuel types.

As a starting point to improve and possibly simplify the analysis of NHOC, we use the standard test method for estimating NHOC, ASTM D3338, as a reference model. This model estimates NHOC based on empirical equations using precise measurements of density at 15°C, aromatic content, distillation curve (average of 10%, 50%, and 90% recovery temperatures), and sulfur content. Consequently, we assess different alternative ML-based modeling approaches to the empirical equations of the ASTM. Furthermore, to address application challenges regarding the difficult-to-measure quantities required by the ASTM model, we study several simplified ML models that rely only on bulk properties as input parameters. As illustrated in Figure 1, the study evaluates the error resulting from various NHOC modeling approaches, using different combinations of input variables, to provide a better understanding of each approach and its potential applications. Additionally, the study aims to identify the optimal set of easy-to-measure input variables most suitable for operational applications.

DATA

The evaluation of the empirical model uses the extensive database of the German Aerospace Center (DLR) [17] to evaluate and compare NHOC estimations methods e.g. ASTM D3338 [12]. The database has been extensively cited in several publications e.g., for detailed composition models [8][18]. It encompasses precise composition and property measurements of jet fuels ranging from conventional fuels (Jet A, Jet A-1, JP-8, etc.) to different alternative fuels (HEFA, ATJ, FT-SPK, etc.) and their blends. This study uses a subset of the DLR database shown in Table 1 to train and evaluate all modeling approaches. Subsequently, the evaluation of the empirical approach assesses NHOC estimations for various fuel types and the resulting errors in comparison to the ASTM D3338 model, with special focus on the effect of using different variations in input parameters. Based on these results, the approach establishes a baseline for dataset-related uncertainty of the model that can be used to evaluate the other modeling approaches.

Table 1. Number of samples per fuel type

Fuel Type	Sample Nr.
Conventional	11.5k
Blends	154
Synthetic	222

ML-MODELS

The evaluation of ML-based adaptations of the ASTM D3338 model relies on different ML regressors (e.g., random forests, gradient boosting) from the Scikit-learn Python Library [19] as alternatives to the fitted empirical equations used in the ASTM method. Due to the potential of ML regressors to capture complex and interdependent relationships, the estimation of jet fuel properties can benefit significantly from the capabilities of ML models. By using the same set of input/output variables as the ASTM method to train and test these models, the adaptation approach emphasizes the ML models' capacity to provide a numerically more efficient and generalized solution to the underlying estimation problem. Such a solution would be less susceptible, for example, to variance not represented in the training dataset. Similar to the evaluation approach, the sensitivity of the resulting models, as well as the correlation to the fuel's composition, are evaluated.

The evaluation of the bulk property-based ML model simplification is based on the formulation of potential ML models that use simple-to-measure parameters, which would, from an application point of view, reduce operational overhead. The approach focuses on models that use only the bulk properties (e.g., density, viscosity, refractive index, electrical conductivity, etc.) and their capacity to model the required quantity, namely NHOC. Similar to the other approaches, the sensitivity of the resulting models and the correlation to the fuel's composition are evaluated. As shown in Figure 2, the study evaluates three modeling approaches for estimating the NHOC:

- The ASTM D3338-based empirical model
- The ML-based adaption of ASTM D3338 model
- Two simplified bulk property-based ML models:
  - NHOC as function of density at 15°C and kinematic viscosity at -20°C
  - NHOC as function of density at 15°C

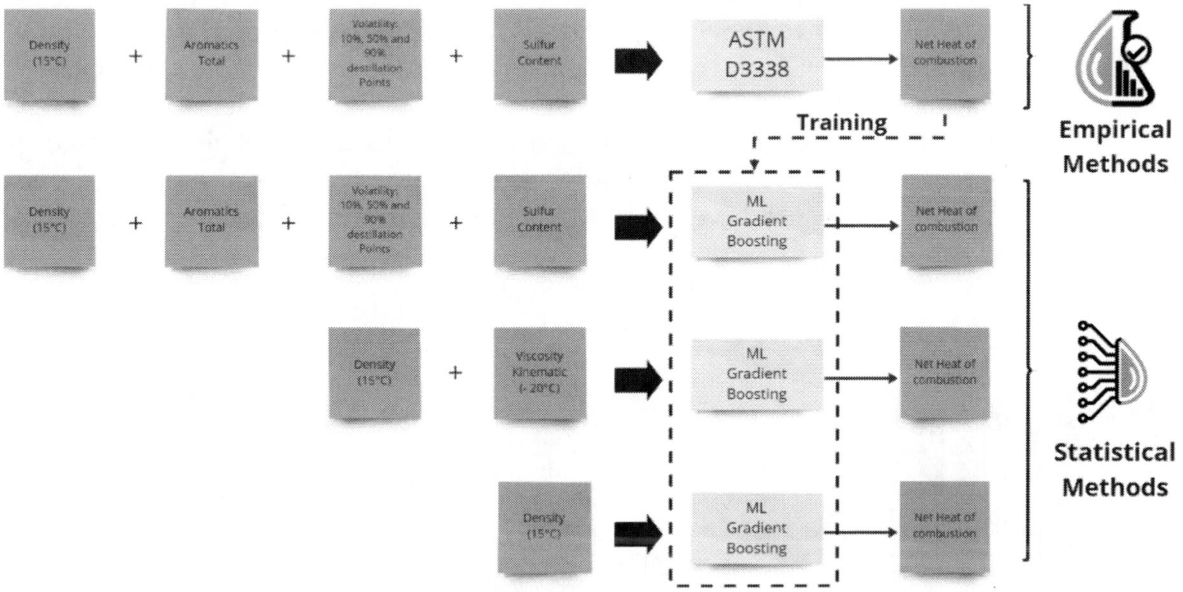

Figure 2: Schematic representation of the model inputs for the ASTM D3338 method and the machine learning models studied in this work. The figure illustrates which values are used for training.

This study focuses on applying gradient boosting machine learning (ML) methods, which provide a balanced approach between decision tree-based methods (e.g., random forests) and other regression techniques (e.g., neural networks). Notably, the chosen method is particularly well-suited for operational usage, as it effectively minimizes bias while handling complex datasets. The model parameters are given in Table 2. A 50/50 split of the dataset was employed for training and testing purposes, in conjunction with 5-fold cross-validation, to yield accurate and realistic results that are directly comparable to operational application scenarios. This approach ensures a robust evaluation of the model by reducing overfitting and providing a comprehensive assessment of its generalization capability. For consistency and comparability, the model values were trained against the ASTM D3338 method, which provided the largest and most widely available dataset across various fuel types.

Table 2. Gradient boosting model parameters

Parameter	Value
Training/testing split	50%/50%
Cross-validation method	5-Fold
Max tree depth	6
Learning rate	0.001

RESULTS

PROPERTY CORRELATION

To evaluate the suitability of different inputs for various modeling approaches, we analyzed the correlation coefficients of fuel properties with the NHOC values estimated by ASTM D3338, as shown in Figure 3. As expected, the hydrogen content has the highest positive correlation with the NHOC value, given its significant contribution to the energy released during the combustion process. Smoke point also correlates positively with NHOC; higher smoke points indicate fewer aromatics and more paraffinic hydrocarbons in the fuel, leading to a higher NHOC. This also explains the negative correlation of NHOC with density, since higher density indicates lower paraffinic content and higher aromatic content. Similarly, the 50% recovery temperature in distillation indicates the presence of heavier, higher-boiling-point hydrocarbons in the fuel, which have lower hydrogen-to-carbon ratios and negatively influence the NHOC.

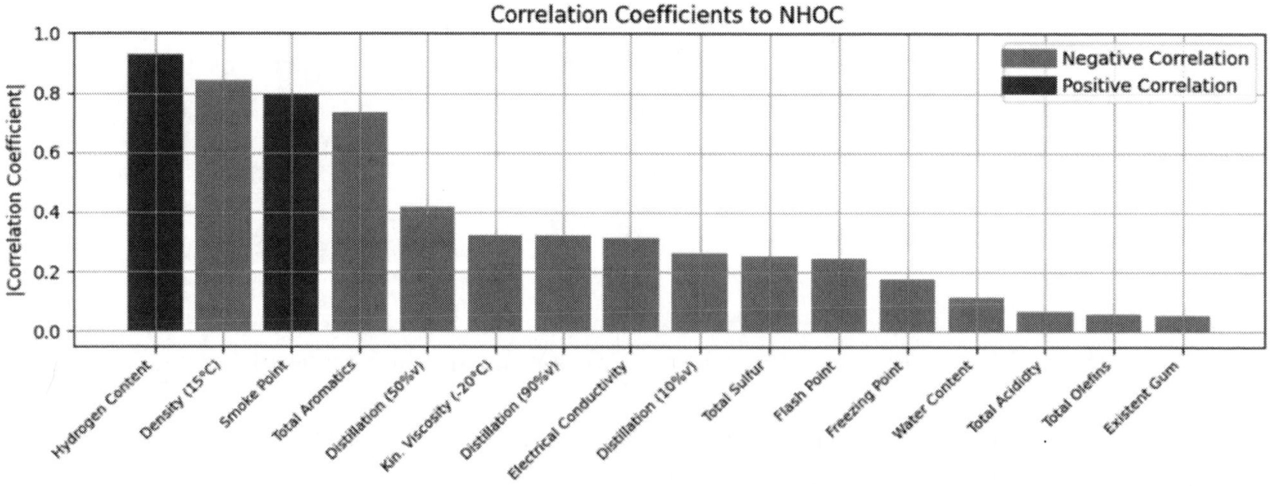

Figure 3: The figure displays the correlation coefficients between various commonly measured fuel properties and the ASTM D3338 NHOC value. Blue bars for positive correlations, red bars for negative.

In terms of the absolute values of the correlation coefficients, the inputs for the ASTM D3338 method (density at 15°C, aromatic content, distillation points) show the strongest correlations, underscoring their significance for modeling NHOC. While density is definitely the best input for a simplified bulk property modeling approach, the next best parameter is the kinematic viscosity at –20°C, which has a comparable correlation value to the 50% recovery temperature in distillation. As shown in Figure 4, kinematic viscosity at –20°C functions as a proxy for the 50% recovery temperature and the presence of heavier hydrocarbons in the fuel. Therefore, including kinematic viscosity at –20°C as an input alongside density can significantly improve the model's accuracy, making it a valuable parameter for simplified modeling approaches.

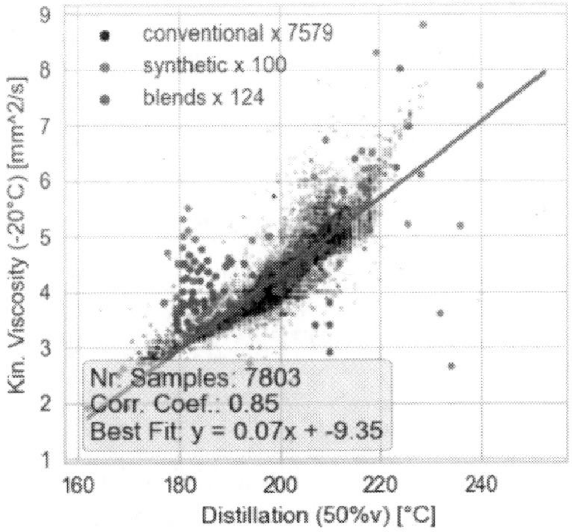

Figure 4: Comparison of kinematic viscosity at –20°C to 50% volume recovery temperature during distillation for different fuel types. The number of samples per fuel type is given in the legend.

NHOC MODELS

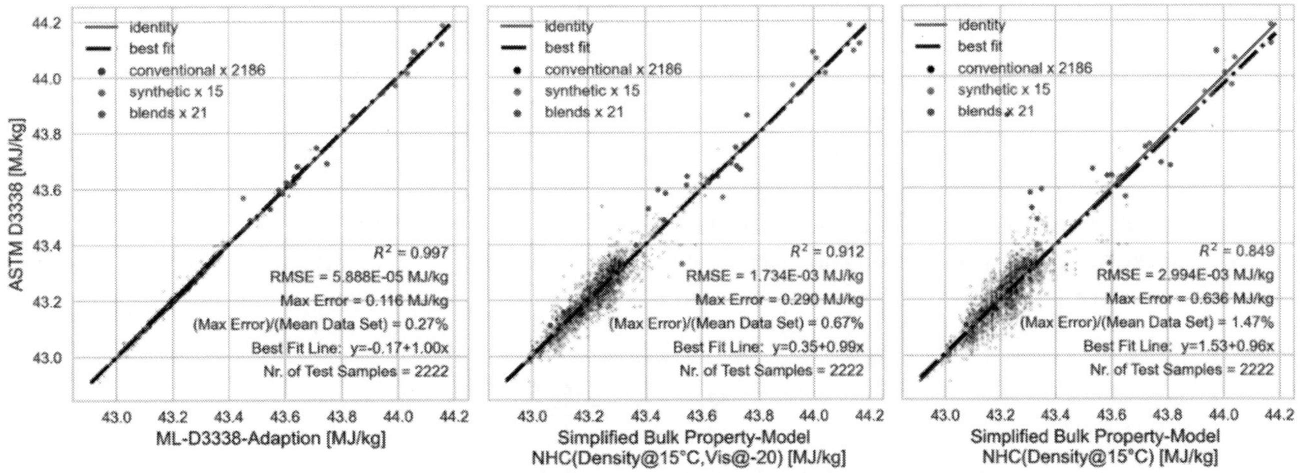

Figure 5: Identity plot comparing the results of ASTM D3338 with three newly formulated statistical models: 1) using the same inputs as ASTM D3338, 2) using density at 15°C and kinematic viscosity at -20°C, and 3) using only density at 15°C. Additionally, the figure presents key performance metrics such as the coefficient of determination (R^2), root mean square error (RMSE), maximum error, and maximum relative error.

As shown in Figure 5, the results demonstrate that using bulk properties density at 15°C and viscosity at –20°C as model inputs yields a very good compromise between accuracy and operational monitoring, as these two properties are in field easy-to-measure model inputs. When the ML model utilizes the same input variables as ASTM D3338, it accurately reproduces the results, achieving an R^2 value approximately equal to 1 and exhibiting the lowest error metrics. The two ML models trained on simplified inputs (one using density at 15°C and viscosity at –20°C, and the other using only density at 15°C) also show very promising results and high R^2 values. The largest scatter in these simplified models can be attributed to blends. The inclusion of kinematic viscosity, despite its relatively low correlation coefficient to NHOC as shown in Figure 3, leads to improved accuracy and error metrics compared to using density alone. This can be explained by the strong correlation of kinematic viscosity with the 50% recovery temperature in the distillation, which is one of the four input parameters in the ASTM D3338 model. These results support the premise of the study that, for a given application and required accuracy, such simplified ML models could be considered.

CONCLUSION

This study demonstrates the potential of alternative modeling approaches for NHOC using ML techniques. The main finding is that simplified modeling approaches can accept some degree of uncertainty in exchange for using simple, easy-to-measure model inputs. The error comparison scheme employed enhances model interpretability and helps mitigate "black box" effects often associated with complex models. The analysis of the error baseline ASTM D3338 and three ML-based models highlights the importance of balancing simplicity and accuracy. Our results suggest that various modeling approaches can be considered to ensure the best estimation accuracy for specific applications.

The findings presented offer a promising use case for alternative modeling of NHOC, which can bolster decision-making in online and inline application scenarios. Specifically, the ML-based adaptation model matched the results of ASTM D3338 when tested on the same dataset, while the simplified ML model based solely on bulk properties (density at 15°C and kinematic viscosity at –20°C) yielded promising results with a root mean square error of 0.002 MJ/kg and a maximum relative error of less than 0.7%. These results suggest that bulk properties and ML-based models can be used to resolve complex quantities such as NHOC for well-defined aviation fuels.

Looking ahead, our research aims to expand the modeling approach to other properties of interest and develop model applications and hardware that can leverage these techniques. Future work will focus on training ML models on ASTM D4809 (Bomb Calorimeter Test Method) using transfer learning, which has the potential to improve estimation accuracy and interpretability. Through this research, we aim to develop more accurate and reliable modeling approaches for NHOC and other aviation fuel properties.

References

1. ASTM D1655, "Standard Specification for Aviation Turbine Fuels."

2. ASTM D7566, "Standard Specification for Aviation Turbine Fuel Containing Synthesized Hydrocarbons."

3. ASTM D4809, "Standard Test Method for Heat of Combustion of Liquid Hydrocarbon Fuels by

Bomb Calorimeter (Precision Method)."

4. ASTM D2386, "Standard Test Method for Freezing Point of Aviation Fuels."
5. ASTM D1319, "Standard Test Method for Hydrocarbon Types in Liquid Petroleum Products by Fluorescent Indicator Adsorption."
6. ASTM D5769, "Standard Test Method for Determination of Benzene, Toluene, and Total Aromatics in Finished Gasolines by Gas Chromatography/Mass Spectrometry."
7. Heyne, J., Ciepiela, L., and Edwards, T. (2021), "Sustainable Aviation Fuel Prescreening Tools and Procedures," *Fuel*, `https://doi.org/10.1016/j.fuel.2020.120004`
8. Hall, C., Pitz, W. J., Whitesides, R., and Smith, A. (2022), "Probabilistic Mean Quantitative Structure–Property Relationship Modeling of Jet Fuel Properties," *Energy & Fuels*, `https://doi.org/10.1021/acs.energyfuels.1c03334`
9. Vozka, P., Maki, K. C., and Heyne, J. (2018), "Impact of HEFA Feedstocks on Fuel Composition and Properties in Blends with Jet A," *Energy & Fuels*, `https://doi.org/10.1021/acs.energyfuels.8b02787`
10. Boehm, R. (2022), "Lower Heating Value of Jet Fuel from Hydrocarbon Class Concentration Data and Thermo-Chemical Reference Data: An Uncertainty Quantification," *Fuel* `https://doi.org/10.1016/j.fuel.2021.122542`
11. Edwards, T. (2023), "Jet Fuel Properties," *United States Air Force Research Laboratory*, AFRL-RQ-WP-TR-2020-0017.
12. ASTM D3338, "Standard Test Method for Estimation of Net Heat of Combustion of Aviation Fuels."
13. Alqaheem, S. and Riazi, M. R. (2017),"Flash Points of Hydrocarbons and Petroleum Products: Prediction and Evaluation of Methods", *Energy & Fuels*
14. ASTM D3343, "Standard Test Method for Estimation of Hydrogen Content of Aviation Fuels."
15. ASTM D976, "Standard Test Method for Calculated Cetane Index of Distillate Fuels."
16. ASTM D4737, "Standard Test Method for Calculated Cetane Index by Four Variable Equation."
17. Bauder, U., Hall, C., Pütz, F., Enderle, B., Le Clercq, P. and Rauch, B. (2022) "An Interactive, Interdisciplinary and Collaborative Digital Platform for the Assessment and Optimization of Jet Fuels," *IASH 2017*
18. Hall, C., Pitz, W. J., Whitesides, R., and Smith, A. (2021), "Predictive Capability Assessment of Probabilistic Machine Learning Models for Density Prediction of Conventional and Synthetic Jet Fuels," *Energy & Fuels*, `https://doi.org/10.1021/acs.energyfuels.0c03779`
19. Pedregosa, F., Varoquaux, G., Gramfort, A., Michel, V., Et al. (2011), "Scikit-learn: Machine Learning in Python," *Journal of Machine Learning Research*

535d Study on Single-Step Hydrogenation of Soybean Oil for Biofuel with ZSM-5 Supported Nickel-Molybdenum Bimetallic Catalysts

Yu-Hsuan Cheng, Bing-Hung Chen, Duu-Jong Lee

In this work, the effect on different alkali treatments of zeolite ZSM-5 for Ni-Mo/ZSM-5 catalysts to the single-step hydrogenation of soybean oil for biofuels was studied. The properties of zeolite ZSM-5 as an acidic support were modified through hydrothermal and alkali treatments, prior to loading Ni-Mo bimetallic catalysts via the incipient wetness impregnation. In our conjecture, Ni serves as H_2 activation centers, diffusing activated hydrogen to other active sites through hydrogen spillover mechanisms, while molybdenum provides oxygen vacancies, enhancing selectivity towards C-O bond cleavage. The metallic oxides were firstly reduced to metallic states under hydrogen atmosphere at 550°C. The catalysts were characterized by XRD, XPS, SEM, TEM, ICP-OES, H_2-TPR, NH_3-TPD, Pyridine-IR, CO-pulse and N_2 physisorption analyzer. Unlike the commercial two-step reaction process, this study aimed to achieve hydrogenation-deoxygenation and hydrogenolysis/isomerization reactions simultaneously in one step.

The hydrotreatment of soybean oil was performed using decalin as the solvent and various catalysts in a semi-batch reactor for 6 h. The products were analyzed quantitatively by GC-FID and qualitatively by FTIR and GC-MS. The objective is to achieve the maximum yield while operating within the reaction pressure fixed at 40 bar of hydrogen, and reaction temperatures from 300°C to 350°C. (Figure 1)

Comparing catalysts with zeolite ZSM-5 supports priorly treated with different alkali concentrations, it's observed that selectivity of liquid hydrocarbon products towards higher carbon numbers increased notably with higher alkali concentrations. To attain more sustainable aviation fuel (SAF) as possible, the focus of this research work was shifted towards products with medium carbon numbers. Despite the burgeoning development of alternative energy sources, the aviation industry, reliant on high energy density fossil liquid fuels for long-distance transport, still necessitates them. Hence, ZSM-5 is subjected to hydrothermal treatment to promote the formation of MCM-41/ZSM-5 composite materials, augmenting catalyst surface area and mesopore volume. This augmentation facilitates the diffusion and reaction of triglycerides, thus improving selectivity towards medium carbon numbers. Finally, recyclability tests of the catalyst probe whether the nickel-molybdenum bimetallic catalysts retain good deoxygenation capabilities after catalysis.

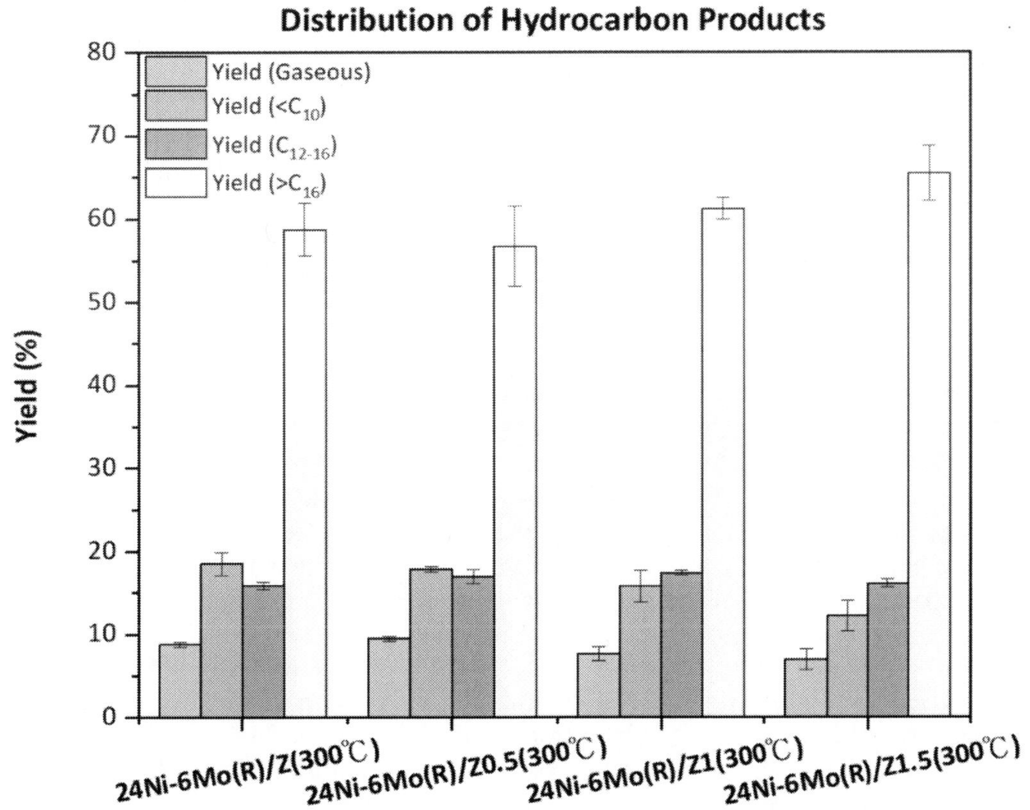

Distribution of Hydrocarbon Products
Yield (%)
0
10
20
30
40
50
60
70
80
Yield (Gaseous)
Yield (<C10)
Yield (C12-16)
Yield (>C16)
24Ni-6Mo(R)/Z(300℃)
24Ni-6Mo(R)/Z0.5(300℃)
24Ni-6Mo(R)/Z1(300℃)
24Ni-6Mo(R)/Z1.5(300℃)

535e Well-to-Wake Cost and Emissions Assessments for Iron Ore Shipping in the Western Australia–East Asia Green Corridor

Christopher Douglas, Santosh Shanbhogue, Haoxiang Lai, Mohammad Ostadi, Ahmed Ghoniem, Guiyan Zang

Because of its status as a heavy emitter responsible for roughly 3% of all global greenhouse gas (GHG) emissions, the maritime sector is facing increasing pressure to decarbonize. As a result of this pressure, the Clydebank Declaration of 2021 endorsed "green shipping corridors" among signatory countries that will stimulate low carbon shipping through policy incentives designed to hedge key stakeholders, catalyze new technologies, develop robust supply chains, assess tradeoffs, and, ultimately, accelerate maritime industry decarbonization.

This study assesses the life-cycle emissions and costs of various proposed decarbonization solutions for the recently designated green shipping corridor for iron ore transport between Western Australia (WA) and East Asia (EA). Through a attributional life-cycle assessment (LCA) and comparative techno-economic analysis (TEA) of the primary near-term solutions for carrying and converting renewable energy in this application, we identify the most promising approaches to significantly abate the well-to-wake greenhouse gas (GHG) emissions of the existing fossil-powered fleet while maintaining the competitiveness of the WA–EA iron ore shipping corridor.

In the analysis, a representative vessel, cargo, and voyage profile is identified from a survey of current iron ore bulk carrier fleet, which is then used to quantitatively compare the levelized cost and GHG emissions of iron ore delivered among conventional and alternative options. The considered alternative energy carriers include "green" hydrogen, ammonia, and methanol fuels synthesized with water electrolysis using wind energy and renewable CO_2 resources in Australia, and the analysis models the costs and GHG emissions of the fuel supply chain throughout the production, transportation, storage, and end-use stages. All considered solutions are designed with emissions treatment to comply with the International Maritime Organization's (IMO) Tier III NO_x regulations.

Our results reveal that renewable ammonia fuel bears the lowest green premium among the considered options, yielding a levelized total cost of ship ownership approximately 44% higher than an equivalent conventional heavy fuel oil (HFO) vessel and approximately 93% lower levelized well-to-wake GHG emissions. In notable contrast to ships carrying lighter cargo, we find that iron ore bulk carrier ships' carrying capacity is not significantly penalized by the reduced energy density of alternative fuels relative to HFO due to the high density of iron ore. Despite the enticing potential for impact, the baseline scenario yields a relatively high carbon abatement cost of $240 USD per tonne of CO_2-e, excluding subsidies and policy incentives. Sensitivity analysis is also presented to generalize these results to a broader range of scenarios. These results underscore the significant economic challenge of maritime decarbonization, where initiatives such as green corridors will be essential to rapidly bridge the gap between climate benefits and economic viability.

535g Electrochemical Control and Modulation of the Flammability of Ionic Liquid Fuels

Prithwish Biswas, Afrida Anis, Michael Zachariah

Flammability and combustion of high energy density liquid propellants are controlled by their volatility. We demonstrate a new concept through which the volatility of a high energy density liquid propellant can be dynamically manipulated enabling one to (a) store a thermally insensitive oxidation resistant non-flammable fuel (b) generate flammable vapor phase species electrochemically by applying a direct-current voltage bias and (c) extinguishing its flame by removing the voltage bias, which stops its volatilization. We show that a thermally stable imidazolium-based energy dense fuel, can be made flammable or non-flammable simply by application or withdrawal of a direct-current bias. This cycle can be repeated as often as desired. In-situ mass spectrometry is used to characterize what species are volatilized during electrolysis. To explore the role of anion we employ 5 different species and show that the rate of electrochemical reaction scales with ion conductivity implying this is a mass transfer controlled process. The estimated energy penalty of the electrochemical activation process is only ~ 4% of the total energy release. This approach presents a paradigm shift, offering the potential to make a 'safe fuel' or alternatively a simple electrochemically driven fuel metering scheme.

535h Synthesis, Characterization, and Performance of Alkyl Levulinate-Diesel Blends

Sampath Karunarathne, Scott Eaton, Ravikant Patil, Sampath Gunukula, Clayton Wheeler

Renewable diesel plays an important role in decarbonizing the transportation sector. It has an increasing demand particularly for sectors where electrification is not practical or economically feasible such as heavy duty-vehicles, rails, highway freight, and maritime transport. Alkyl levulinates (ALs), which are derivatives of levulinic acid, have garnered significant interest as promising bio-based fuel additives and diesel blend components. In contrast to biodiesel, the components of ALs, levulinic acid and alcohols, can be sustainably produced from various sources of cellulosic biomass, including non-food sources such as wood waste, grass, and organic waste, without competing with food production.

Most of the previous studies have focused on short-chain levulinates (e.g., methyl and ethyl levulinates), which exhibit less favorable fuel properties, for example, poor miscibility, low heating value, and high water solubility. However, long-chain ALs extend the alkyl chain of the levulinate ester, thereby improving the energy density, miscibility, and hydrophobicity of the fuel. Therefore, investigating the potential of both short- and long-chain ALs as diesel blend components is worthwhile.

In this study, synthesis and characterization of short- and long-chain ALs and the performance of alkyl levulinate-diesel blends as fuel blend component are investigated. The ALs were produced via the catalytic esterification of levulinic acid with alcohol using an acid catalyst. A heterogeneous catalyst such as the ion-exchange resin (e.g., Amberlyst-15) is used for the esterification reactions because of the high efficiency of catalyst recovery and product separation. The effect of reaction conditions (reaction temperature, reaction time, catalyst loading, and molar ratio of levulinic acid to alcohol, etc.) on conversion and yield are investigated. The effect of varying alkyl chain lengths in ALs on fuel properties is also studied. Both short-chain (methanol and ethanol) and long-chain (butanol, hexanol, octanol, etc.) alcohols are used to produce ALs with different carbon chain lengths. ALs are recovered from the reaction mixture via distillation and refined to a high-purity final product. Chemical composition and purity of the products are determined by gas chromatography (GC-MS and GC-FID), FTIR, and Karl Fischer titration methods.

Subsequently, ALs are blended with Ultra Low-Sulfur Diesel (ULSD) in different blend ratios (5-20 vol. % blends) and analyzed for physicochemical properties such as solubility, density, flash point, viscosity, heat of combustion, hydrogen content, cetane number, etc. according to standard ASTM methods. Combustion dynamics and emission testing are conducted using a Kohler KD440 single-cylinder diesel engine in genset configuration to investigate the effect of ALs on engine performance, durability, and emission characteristics.

535i Low-Temperature Fabrication of Dense $Ba(Zr,Y)O_3$ Coating Via Sol-Gel Processes for Metal-Supported Protonic Ceramic Fuel Cells

Hyunghoon Song, Jaeseok Lee, Joongmyeon Bae

Metal-supported solid oxide fuel cells (MS-SOFCs), a form factor that replaces the ceramic support of conventional anode-supported SOFCs with metal, offer several advantages over ceramic-supported SOFCs, such as enhanced ruggedness against mechanical shock, rapid thermal transients, and redox cycling. Meanwhile, protonic ceramic fuel cells (PCFCs) have attracted attention due to their lower proton-conducting activation energy compared to oxygen ion-conducting activation energy at intermediate temperatures. To further lower the operating temperature of MS-SOFCs, one potential solution is to fabricate metal-supported cells with proton-conducting electrolytes, such as barium zirconate-cerate-based materials. However, using the conventional co-sintering approach is not possible because the required sintering temperature for the densification of the proton-conducting ceramic electrolyte is too high to sustain the porous microstructure of the metal support. Therefore, low-temperature film deposition processes, such as physical vapor deposition or sol-gel coating, are required to fabricate the metal-supported protonic ceramic fuel cells (MS-PCFCs). Additionally, the good step coverage of sol-gel coating is advantageous for depositing thin and gas-tight electrolytes on porous anode surfaces.

In this work, a sol-gel coating process is developed for dense $BaZr_{0.8}Y_{0.2}O_3$ (BZY) coatings at low-temperatures (< 1000 °C) on a porous anode surface. A precipitation-free sol is synthesized by mixing solvent, chelating agent, metal acetates, and alkoxides in a specific order. The sol is spin-coated on a silicon wafer and the porous anode surface of anode-supported or metal-supported cells for X-ray diffraction analysis and microstructure characterizations. It is shown that a crystalline BZY film without secondary phases can be formed by heat treatment above 800 °C. Furthermore, cross-sectional SEM analysis reveals that a thin and crack-free BZY film can be successfully deposited on a porous anode surface at heat treatment temperatures below 1000 °C. The developed coating process can be one of the feasible solutions for successful MS-PCFCs fabrication.

328a Techno-Economic Assessment of Biofuel Production from the Hydrothermal Liquefaction of Macroalgae

Micheal Asama, Fernando Resende, Aaditya Khanal

The hydrothermal liquefaction (HTL) of macroalgae has garnered considerable attention as a promising environmentally sustainable method for producing renewable fuels and other high-value chemicals. In this comprehensive study, we conducted an extensive techno-economic analysis (TEA) and life cycle assessment to evaluate the feasibility, economic viability, and environmental impact of developing a commercial-scale HTL process for converting macroalgae into valuable products. We leveraged experimental data to develop a TEA model for the HTL of 25 metric tons per hour (MT/hr) of macroalgae to produce 5.26MT/hr of biofuel within the gasoline and diesel range fuel. Two cases were evaluated: a conventional case with HTL experimental testing results underlining the major design basis, and an intensified case considering future improvements for a commercial plant with advanced technologies. Process simulation and cost estimates were conducted to evaluate the total project investment and operating cost to 221M$ and 155M$/year, respectively. The annual production rate for the final biocrude was estimated to be 22.5 and 39.9 million gallons of gasoline equivalent (GGE) for the conventional and the intensified cases, respectively. The minimum fuel selling price (MFSP) based on the assumption of a 10% internal rate of return and 30 years of plant life was estimated to be $11/GGE for the conventional case and $8/GGE for the intensified case. We performed a sensitivity analysis to highlight the key variables influencing the target scenario, underscoring the cost uncertainties associated with the assumed variations in selected parameters. Our results showed that product separation strongly influences the MFSP, reducing production costs when moving from the conventional to the intensified case.

328b Corn Stover to High Value Carbon for Energy Storage

Khang Huynh, Bharathkiran Maddipudi, Anuradha Shende, Rajesh Shende

Worldwide, corn stover is generated in significant amounts. Using the integrated hydrothermal processing approach, corn stover can be converted into high value-added products such as bio-oil, hydrochar, and aqueous coproducts containing carboxylic acids and phenolics. Hydrochar is known to be porous that can be further activated by chemical/thermal methods to achieve high surface area carbon. In this investigation, hydrochar is activated with potassium hydroxide solution and heated to different temperatures, and characterized for specific surface area, pore size, pore volume, elemental composition, and surface functional groups. Energy storage performance of the activated carbons is analyzed in an asymmetric supercapacitor (ASC) configuration. ASCs possess higher cycle life as compared to batteries and superior specific capacitance as compared to electric double-layer capacitors (EDLCs) due to their charge storage mechanism by Faradaic reaction between the electrode and electrolyte. The process of hydrochar derivatization and activation is scaled-up at 1 tpd scale using ASPEN Plus modeling. Mass and energy balances are studied. Technoeconomic feasibility is determined at different production scales while considering costs associated with raw material, equipment, utilities, wastewater treatment and revenue from the carbon. LCA is also performed to understand the GHG emission for specific system boundary. Hydrochar production, carbon for energy storage, performance evaluation of asymmetric supercapacitor, scale-up, mass and energy balances, ASPEN modeling and TEA/LCA will be presented.

328c Techno-Economic Assessment of Sustainable Biolubricant Production from Lignin and Waste Cooking Oil

Elena Barbera, Andrea Guarise, Randy Maglinao, Sandeep Kumar

Lignin accounts for 15-30% of lignocellulosic biomass weight and 40% of the biomass heat energy. Today, lignin from lignocellulosic biorefineries is underutilized because of the absence of a high-value product from wet unhydrolyzed solids (UHS), a waste stream that has very little value. Finding ways to valorize this byproduct is crucial to enhance the economic viability and sustainability of biofuels production.

In this work, a process is developed to produce a sustainable biolubricant starting from waste cooking oil (WCO) and lignin from UHS. The development of sustainable and environmentally friendly lubricants has in fact become an important challenge, and their market has increased significantly in recent years.

The proposed process consists of three steps: (1) Hydrothermal liquefaction (HTL) of lignin, to depolymerize it into monomeric phenolic compounds, collected into a bio-oil; (2) Hydrodeoxygenation (HDO) of the bio-oil to replace the hydroxyl group of phenolic compounds with hydrogen, and thus produce aromatic hydrocarbons; (3) alkylation of unsaturated FAMEs (in particular from transesterification of WCO) with the aromatic hydrocarbons obtained from lignin degradation.

Experiments were first carried out to find out the optimal operating conditions that maximize the reaction yields of the three process steps. In particular, the HTL step was conducted using an extremely efficient induction heating system that allowed reaching the reaction temperature within 2-3 min. The optimal operating conditions that allowed to maximize the bio-oil yield were found to be 320 °C at 1 min residence time, which resulted in the highest selectivity of phenol (52 wt%).

The hydrodeoxygenation of a mixture of phenolic compounds was performed at 250°C, utilizing alkane solvents. To improve both yields and selectivity toward the formation of aromatic hydrocarbons, a multi-step approach was adopted. This strategy yielded a conversion rate of 81 wt%, with 58 wt% selectivity. In the final step, the alkylation of the aromatic hydrocarbons using unsaturated FAMEs occurred at 210°C for 4 hours, employing an acid-treated montmorillonite catalysts under argon gas. The resulting product contains 19 wt% alkylated fatty acid methyl esters.

Based on the experimental data collected, a techno-economic analysis was conducted to verify the scalability and economic viability of the proposed process when integrated in an already existing lignocellulosic biorefinery. In particular, capital and operating expenditures were evaluated to determine the minimum selling price of the bioethanol produced by the integrated biorefinery to that of the conventional one.

328d Integrated Processing of Agriculture Residue: Techno-Economic Analysis and Life Cycle Assessment

Bharathkiran Maddipudi, Khang Huynh, Zoe Fickbohm, Anuradha Shende, John E. Aston, Sandeep Kumar, Ram Gupta, Rajesh Shende

Abundantly available agriculture residue (e.g. corn stover) can be subjected to integrated biochemical and hydrothermal processing to generate fuel and value-added products. Although conventional biochemical processing for fuel production generates solid waste, it can be hydrothermally treated further to generate products such as oil and hydrochar. Subsequent valorization of oil by hydrodeoxygenation and hydrochar activation by chemical/thermal methods will produce high density fuel and high value carbon, respectively. Aqueous coproduct (ACP), which is generally recovered after hydrothermal processing generally contains carboxylic acids, phenolics, and other compounds. ACP can be recycled to minimize wastewater production or treated to achieve separation of carboxylic acids and phenolics. In this study, hydrothermal liquefaction (HTL) is employed to treat corn stover and the unhydrolyzed solids (UHS) recovered from biochemical processing and the process is optimized to achieve value-added products for circular bioeconomy. Different reaction parameters are investigated to achieve optimum yields of different products. ASPEN plus modeling is performed to scale-up the integrated process, and techno economic analysis (TEA) and life cycle analysis (LCA) are performed. GREET software is used to assess LCA of integrated bioprocessing of corn stover and UHS. Flow and cost summary with cash flow analysis, sensitivity analysis of change in internal rate of return (IRR) and change in net present value (NPV) will be presented in addition to reduction in GHG emission.

328e Opportunities of Bio-Naphtha Beyond Gasoline Blend: A Comprehensive Techno-Economic and Life-Cycle Analysis

Shuang Xu, Yuan Jiang, Dilara Goreke, Huamin Wang

Biofuels, especially sustainable aviation fuel (SAF) and renewable diesel (RD), play an important role in decarbonizing various hard-to-electrify sectors [1, 2], and the long-term strategy of the Unites States [3] targeting 50-52% greenhouse gas (GHG) emission reduction by 2030, and net-zero by 2050. However, the production of SAF and RD often results in a substantial byproduct, bio-naphtha, with less market demand in a low carbon economy. Therefore, it is crucial to systematically investigate potential utilization pathways for bio-naphtha, considering its characteristics and the variations arising from different biomass/waste feedstocks and conversion technologies.

This study conducts detailed analysis of bio-naphtha derived from four distinct biofuel technologies, Fischer-Tropsch (FT), hydrothermal liquefaction (HTL), catalytic fast pyrolysis (CFP), and hydroprocessed ester and fatty acids (HEFA), and three potential utilization approaches, steam cracking for producing either polymer grade olefins, or mixed olefins for SAF production, and catalytic reforming for producing green hydrogen and aromatics. Kinetic-based reactor models are developed in Aspen Plus or Aspen HYSYS to evaluate the impact of feedstock composition on reactor performance and product selectivity. Feedstock characterization is collected from either inhouse experiment or literature. Life cycle analysis (LCA) and techno-economic analysis (TEA) are conducted based on plant-wide process model to assess the environmental and economic impacts of different bio-naphtha and utilization technologies. Ultimately, the LCA and TEA results provide guidance for the decision-making regarding the most suitable pathways for bio-naphtha utilization, thereby contributing to the advancement of sustainable biofuel production practices.

References

[1] https://www.energy.gov/eere/bioenergy/sustainable-aviation-fuel-grand-challenge

[2] https://afdc.energy.gov/fuels/renewable_diesel.html

[3] https://www.whitehouse.gov/wp-content/uploads/2021/10/us-long-term-strategy.pdf

328f Surfactant Assisted Hydrothermal Liquefaction for Oil and Hydrochar

Bharathkiran Maddipudi, Khang Huynh, Zoe Fickbohm, Anuradha Shende, Rajesh Shende

This study reports surfactant assisted hydrothermal liquefaction of corn stover. During hydrothermal processing, a surfactant may act as a barrier and create conditions to yield pseudo lignin that can translate into higher oil production. The role of surfactant is not yet well understood. This investigation is aimed at developing fundamental understanding of different surfactants during HTL. Catalytic HTL (CHTL) of corn stover is performed at 225 °C to 300 °C and 800- 1300 psi (system autogenous pressure) in the presence of anionic, cationic, and non-ionic surfactants. Effects of different reaction parameters such as temperature, initial nitrogen pressure, biomass to water ratio, surfactant concentration, and catalyst loading on product distribution are evaluated. Presence of a surfactant can change the wettability, reduce the hydrophilicity, and increase interfacial surface area, thus influence hydrochar characteristics as well as distribution of liquified products, such as C_1-C_3 carboxylic acids. The aqueous biocrude is characterized by TOC analyzer and gas-chromatography mass spectrometry (GC-MS) to understand extent of liquefied carbon and oxygenated hydrocarbons, respectively. Both hydrochar and HBO are analyzed by the elemental analyzer, and HHVs are calculated based on C, H, N and O values. Hydrochar is characterized by BET surface area analyzer and FTIR. Results obtained with the surfactant assisted CHTL of corn stover, mass and energy balance will be presented.

This study reports surfactant assisted hydrothermal liquefaction of corn stover. During hydrothermal processing, a surfactant may act as a barrier and create conditions to yield pseudo lignin that can translate into higher oil production. The role of surfactant is not yet well understood. This investigation is aimed at developing fundamental understanding of different surfactants during HTL. Catalytic HTL (CHTL) of corn stover is performed at 225
°C to 300 °C and 800- 1300 psi (system autogenous pressure) in the presence of anionic, cationic, and non-ionic surfactants. Effects of different reaction parameters such as temperature, initial nitrogen pressure, biomass to water ratio, surfactant concentration, and catalyst loading on product distribution are evaluated. Presence of a surfactant can change the wettability, reduce the hydrophilicity, and increase interfacial surface area, thus influence hydrochar characteristics as well as distribution of liquified products, such as C_1-C_3 carboxylic acids. The aqueous biocrude is characterized by TOC analyzer and gas-chromatography mass spectrometry (GC-MS) to understand extent of liquefied carbon and oxygenated hydrocarbons, respectively. Both hydrochar and HBO are analyzed by the elemental analyzer, and HHVs are calculated based on C, H, N and O values.
Hydrochar is characterized by BET surface area analyzer and FTIR. Results obtained with the surfactant assisted CHTL of corn stover, mass and energy balance will be presented.

328g The Biofine Process: Carbon-Negative Fuels and Chemicals - a Status Report

Stephen Fitzpatrick

The Biofine process allows conversion of a wide range of lignocellulosic or cellulosic wastes into fuels and commodity chemicals with huge markets. The process is based on the acid hydrolysis of the carbohydrate portions of the feedstock (cellulose and hemicellulose) under relatively mild conditions of dilute sulfuric acid and moderate temperature that dissolve and convert the carbohydrates but allow lignin in the feedstock to pass through the reaction relatively unchanged. Carbohydrates are converted simultaneously to the platform chemicals, levulinic acid, formic acid, and furfural. High simultaneous yields of the products are obtained by virtue of the stable mixing regimes established within a continuous back-mixed, steam-stripped, reaction vessel.

The biofuel of primary interest is ethyl levulinate (EL), the ethanol ester of levulinic acid suitable for both heavy and light fuel formulations. The initial market for EL will be commercial and home heating oil. The byproducts are levulinic acid, a versatile platform chemical that can be readily converted into monomers such as succinic acid, acrylic acid and 3-HPA, sodium formate, an environmentally benign non-corrosive airport and bridge deicer and a high-energy content, non-bio-degradable carbonaceous char consisting of unconverted lignin and reaction residue. Independent carbon dioxide life cycle analyses (LCA) have been carried out on production of both the fuel and the biochar using the "GREET" model developed at Argonne National Laboratory. These conclude that for both products the commercial scale process is carbon negative: The displacement of fossil feedstocks, avoidance of emissions from cellulosic waste and sequestration of carbon in the biochar residue results in the *reduction* of atmospheric carbon dioxide by up to 20 Kg per gigajoule of biofuel.

The company in recent years has made major strides towards commercialization of the technology and is planning the start of construction of its first commercial plant in Q4, 2024. The process technology has been proven at large pilot scale for several years at the University of Maine (Orono) and has been validated as ready for scale-up in an independent peer review by an international independent engineering consulting firm. The company has entered a multi-year offtake contract with a large national fuel distributor and has verified the applicability of EPA-managed D-7 RINS (renewable Information numbers) for its biofuel product. It has also received a high degree of market interest in the deicer and the carbon sequestrant.

Biofine has also secured a site for construction of the first commercial plant at an abandoned pulp and paper mill in Lincoln, ME and is currently working with a large international EPC company carrying out detailed process design. This first plant will process a total of over 40,000 dry metric tons of waste wood and post-consumer fiber per year and produce over 10 million liters of fuel.

The economics of the process make it highly profitable even at small commercial scale. The primary markets being served, biofuel (EL), carbon sequestration (biochar) and runway and bridge deicers (sodium formate) are sufficiently large to absorb the output from dozens of large scale plants both domestic and international. The feedstock is extremely plentiful being derived in rural settings from forest slash and thinning and in urban settings from municipal waste. In the state of Maine this represents a classic example of economic circularity in either rural or urban settings where readily available waste feedstock can be converted into products such as heating oil and deicer with large market demand in the local region while contributing significantly to reduction of atmospheric carbon.

This presentation will discuss the process technology and its scale-up, products, feedstocks and economic projections.

379a Modular Processing of Flare Gas for Hydrogen and Carbon Nanofibers

Jessica Hauck, Kent J. Warren, Samantha Harshberger, Boning Wang, Andrew Broerman, Theodore Champ, Mija H. Hubler, Alan W. Weimer

In the United States, the "shale boom" and wide implementation of hydraulic fracturing has dramatically reduced the price of natural gas. As a result, oil producers increasingly choose to flare associated natural gas rather than sell it, as investing in expensive transportation infrastructure, such as pipelines, is less economically prudent. Moreover, tight oil wells exhaust quickly as production from a single well depletes rapidly, thus imposing difficulties in predicting the optimal pipeline location[1]. A profitable alternative to harnessing this waste of natural resources is needed.

A one-step Catalytic-Chemical Vapor Deposition (CCVD) process to produce carbon nanofibers (CNFs), carbon nanoparticles (CNPs), and hydrogen is a financially promising repurposing of the flared natural gas. An economically viable CCVD process largely depends on the production of a cheap catalyst. Particle Atomic Layer Deposition (ALD) enables the formation of highly dispersed metal nanocatalysts on a given support. The highly dispersed low-cost catalyst is essential as this process contains a *sacrificial catalyst*; the catalyst, support, CNFs and CNPs combine as one product to improve the strength and durability of concrete. The sacrificial catalyst circumvents the well-known disadvantages of rapid catalyst deactivation in methane thermal cracking[2] and avoids the costly step of catalyst/product separation. To combat the challenges of rapid well depletion, this process will be scaled up to a modular unit that can be transferred between hydraulic fracturing sites. If each flare site in the United States transitioned to this modular CCVD process, over 30 million metric tons of CO_2 emissions per annum could be eliminated[3].

In this work, transition metals are deposited onto a silica fume support via ALD. CNFs and CNPs are then produced via CCVD of methane with the ALD nanocatalyst. The carbon product is evaluated for its improvements to concrete durability via various compression and deformation tests. The carbon product evaluation and empirical reaction data inform both the modular unit design and technoeconomic analysis.

[1] Energy, B. *The Williston Basin: Greasing the Gears for Growth in North Dakota*; BENTEK Energy: 2012, 2012

[2] Abanades, A.; Ruiz, E.; Ferruelo, E. M.; Hernandez, F.; Cabanillas, A.; Martinez-Val, J. M.; Rubio, J. A.; Lopez, C.; Gavela, R.; Barrera, G.; Rubbia, C.; Salmieri, D.; Rodilla, E.; Gutierrez, D., Experimental analysis of direct thermal methane cracking. *International Journal of Hydrogen Energy* 2011, *36* (20), 12877-12886.

[3] U.S. Energy Information Administration. Natural Gas Gross Withdrawals and Production.

http://www.eia.gov/dnav/ng/ng_prod_sum_a_epg0_vgv_mmcf_a.htm (accessed April, 2024)

379b Optimization of Process Parameters in Resin-Wafer Electrodeionization for Enhanced Gluconic Acid Recovery from Hemicellulose Hydrolysate

Olabode Akindolani, Jamie Hestekin

There has been significant interest in utilizing renewable biomass as a primary feedstock in pursuing sustainable and eco-conscious chemical processes. Hemicellulose hydrolysate is a byproduct of biomass processing, representing a promising resource for producing gluconic acid, a versatile compound with various industrial applications spanning pharmaceuticals, food additives, and industrial processes. In current biomass processing practices, gluconic acid recovery from hemicellulose hydrolysate remains inefficient due to suboptimal purification techniques. Conventional methods suffer from low yields, energy inefficiency, and impurity issues, limiting gluconic acid production's economic viability and sustainability.

In recent years, resin-wafer electrodeionization (RW-EDI) has emerged as a promising separation technology for recovering gluconic acid from complex aqueous streams. This process involves using ion-exchange resins and an electric field to facilitate the migration of ions across a selective membrane, thereby separating and concentrating target molecules such as gluconic acid. RW-EDI offers several advantages over traditional separation methods, including continuous operation, high purity, and reduced chemical usage. Despite its potential, the efficiency and cost-effectiveness of RW-EDI for gluconic acid recovery depend heavily on the optimization of process parameters. Factors such as current density, flow rate, pH, temperature, resin type, and membrane characteristics significantly influence the performance of the RW-EDI system. However, the complex interplay between these parameters and their optimal values for maximum gluconic acid recovery still needs to be explored.

Due to its simplicity and cost-effectiveness, ion-exchange resins are typically the primary method for acidifying organic salts. However, bipolar membranes (BM) favor a more environmentally friendly approach for converting acid salts into their acidic forms. This is because BM technology produces minimal chemical effluents, thus eliminating the need for cumbersome regeneration steps associated with ion-exchange technologies. The process of producing organic acids through BM has been extensively researched. This method relies on the capability of BMs to split water into hydrogen and hydroxyl ions at the membrane interface. Consequently, the acid salts dissociate, forming acids and bases with the H^+ and OH^- ions generated by the bipolar membrane.

The choice of incorporating a bipolar membrane within the RW-EDI system emerges as a pivotal component in facilitating the selective separation of ionic species, thereby enhancing the efficiency of gluconic acid recovery. By harnessing the unique properties of the bipolar membrane, we mitigated ion crossover and improved the separation of gluconic acid from the complex matrix of hemicellulose hydrolysate while promoting sustainability.

This work used high-pressure liquid chromatography (HPLC) to analyze concentrations of gluconic acid in the diluate and concentrate streams. By identifying critical process parameters such as flow rate, current density, applied voltage, ion exchange resin type, and hydrolysate concentration, the results showed that resin wafer electrodeionization improved the recovery of gluconic acid from hemicellulose hydrolysate, by maximizing efficiency and minimizing energy consumption.

379c Application of Ethanol Reforming in Internal Combustion Engines: Evaluation of Efficiency

Daniel P. P. Rodrigues, Sílvio S. Sumioshi, Gustavo V. Olivieri, Ricardo Torres

The development of new technologies and methods to reduce pollutant emissions is a key focus of the Brazilian federal program ROTA 2030, enabling the synergy between industry and academia to advance and innovate the automotive sector. Applied research in internal combustion engines specifically aims to address the needs of the Brazilian domestic market by promoting the employment of biofuels and enhancing energy efficiency. The integration of a reformer catalyst into the engine is strategic for elevating efficiency standards and for benefiting the current structure of the Brazilian fuel supply. This catalyst transforms ethanol and water into hydrogen and byproducts, which become a proportion with the fresh charge into the engine intake. This process enhances fuel conversion efficiency through chemically enhanced combustion progress. Additionally, the required driving energy of the catalyst reactions and the ideal working temperature are sourced from the residual enthalpy of the exhaust engine gases, thus increasing the exergy of the system. Computer simulations using the AVL CFD software were conducted to predict exhaust gas temperatures and thermodynamic availability under regular engine load variations. The powertrain model includes a detailed arrangement, including a total thermal transient engine, rotational mechanical driveline, resistance forces, control, automation, and a predictive flame kernel combustion model running on the FTP75 cycle. To refine the combustion model, prior simulations with ANSYS CHEMKIN were performed to obtain the laminar flame speeds of the mixtures.

The simulation results guided the scaling of the prototype, aiming to achieve a 10% reduction in brake-specific fuel consumption. The present study was supported by FUNDEP – Program ROTA 2030, with process grant #201798*08.

379d An Updated Chemical Kinetic Model of NH_3 and NH_3/H_2 Oxidation Using Reaction Mechanism Generator (RMG)

Chuangchuang Cao, Alon Grinberg Dana, Kfir Kaplan, Angiras Menon, Kevin Spiekermann, William H. Green**

Due to the pressing need for global carbon neutrality and the mitigation of climate change, ammonia has emerged as a viable renewable energy carrier and a potential substitute for traditional fossil fuels in propulsion and power generation systems. However, owing to the suboptimal ignition characteristics and high NOx emissions, its blending with H_2 has been found to be efficient for technical applications. Although the chemical reaction mechanisms of pure NH_3 have been explored widely, large discrepancies and ambiguous sources still persist in thermo-kinetic parameters across literature models, hindering comprehensive model prediction for experimental measurements over a wide range of conditions. Our study compares all species thermodynamic properties and reactions which rate coefficients are in significant disagreement in the recent literature models. In particular, the difference in $\Delta G^0{}_f$ of several N-species such as N_2H_3 at 1000 K can exceed 50 kJ/mol, and the rate constant of primary fuel oxidation reaction (NH_3+O_2=NH_2+HO_2) exhibits a discrepancy of up to ~ 4 orders of magnitude at 1000 K between literature models published in the last five years. After rigorous evaluation on the thermo-kinetic parameters, we provide the available recommended values or conduct theoretical calculations at the CCSD(T)-F12/cc-pVTZ-F12 level of theory. Based on the updated kinetics and thermodynamics database, we use the Reaction Mechanism Generator (RMG) software to generate a new comprehensive NH_3 and NH_3/H_2 oxidation model, which is thoroughly benchmarked against literature experimental observations of pure NH_3 and NH_3/H_2 mixtures. The generated kinetic model has better or comparable prediction performance without any modifications nor fittings in the thermo-kinetic parameters, while the prediction of NOx and N_2O formation still needs to be refined. Such a model facilities a reasonable interpretation for the reported data, revealing that H_2 addition can boost NH_3 reactivity at different conditions. From the chemical effect perspective, this feature is primarily due to the enhanced production of reactive radicals such as H, OH and HO_2 as H_2 is added. However, H_2 addition to NH_3 oxidation may elevate NOx emissions compared to pure NH_3 oxidation, which can be attributed to the more competitive NO formation than DeNOx-related reactions involving the aforementioned reactive radicals.

Process design of oxy-fuel combustion combined cycles for carbon capture using cold energy of LNG regasification

Yu-Hsun Tsai[a] and Wei Wu[a*]

[a] Department of Chemical Engineering
National Cheng Kung University, Tainan 701401 Taiwan

*Corresponding Author's E-mail: weiwu@gs.ncku.edu.tw

Abstract

Natural gas (NG) serves as a transitional energy source, characterized by lower carbon emissions compared to coal and oil. The CO2 emissions from burning natural gas are approximately half of those from coal, and it releases fewer pollutants. This positions natural gas as a crucial player in the global transition towards a low-carbon economy.

Imported liquefied natural gas (LNG) must be regasified before use. The cold energy released during the LNG regasification process has significant utilization potential, yet in many facilities, this cold energy is not effectively utilized and is wasted. By designing and implementing more efficient cold energy recovery systems, as well as promoting synergy and integration between industrial facilities, the energy efficiency of the LNG regasification process can be significantly improved, reducing environmental impact.

This study aims to design and evaluate an integrated system that utilizes the cold energy from LNG regasification, oxy-fuel combustion, and carbon capture technologies to achieve the goal of zero carbon emissions by 2050. With the growing global energy demand and increasing environmental pressure, governments worldwide are enacting stringent emission reduction policies to promote the development of low-carbon technologies. Natural gas, as a transitional energy source, is becoming an essential component in the emission reduction process due to its relatively low carbon emissions and high energy efficiency. The development of LNG technology further increases the availability of natural gas, and the cold energy released during the LNG regasification process has high utilization value.

This study employs a cryogenic air separation unit (CASU) to produce pure oxygen, with the air feed pre-cooled using the cold energy from LNG regasification to reduce energy consumption. The natural gas is then used as fuel for the gas turbine, undergoing oxy-fuel combustion with pure oxygen. The heat recovery boiler heats water to generate steam, driving the steam turbine. Finally, the cold energy from LNG is used to condense flue gas, separating out carbon dioxide. Compared to conventional NG power plants, the proposed LNG regasification-based power plant not only produces high-purity CO2 (above 99.9%) as a source for value-added chemical products, but also has overall energy efficiency superior to traditional NG power plants.

The plant model was constructed using Aspen HYSYS and compared with other literature to validate its accuracy. The NG combined cycle power plant based on the LNG regasification process is shown in **Fig.1**.

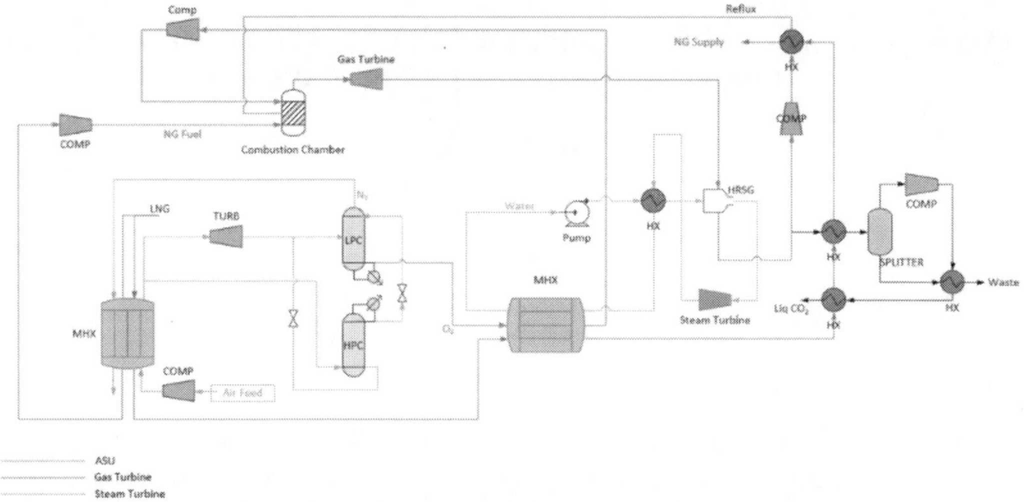

Fig.1 NG combined cycle power plant using the LNG regasification process

The flowsheet of the cryogenic air separation unit (CASU) is shown in **Fig.2**. A double-column configuration is selected, which is the most widely used design. The column configuration of the CASU is similar to that in Yeoh and Hui (2022), except that no additional vapor distillate is drawn from the high-pressure column.

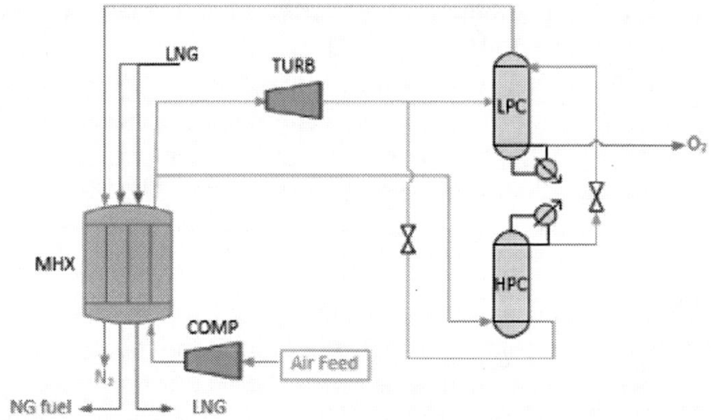

Fig.2 Flowsheet of the CASU

Fig.3 illustrates the oxy-fuel combustion NG combined cycle considering the carbon capture process. In the oxy-fuel combustion system, part of the flue gas is compressed to the same pressure as the combustion chamber, cooled, and then recirculated as a diluent to control the temperature. The remaining flue gas is cooled to separate out carbon dioxide with a concentration greater than 99.9%, which is then compressed and condensed as a valuable product. The cooling required in this process is provided by the cold energy from LNG regasification.

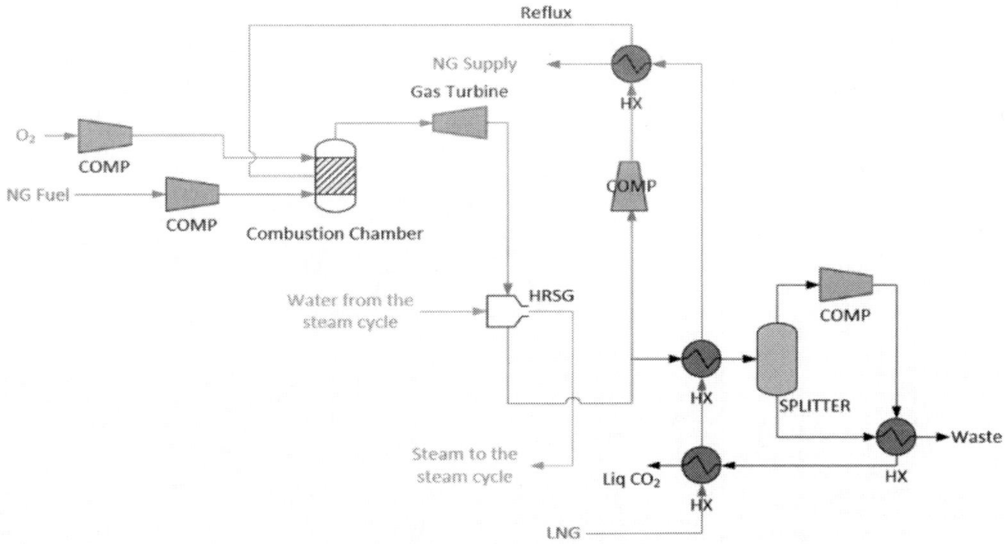

Fig.3 Flowsheet of the GT

Fig.4 shows the single-pressure steam turbine cycle, where high-pressure water is heated in heat exchangers to superheated steam. The high-temperature and high-pressure steam expands in the steam turbine to generate power.

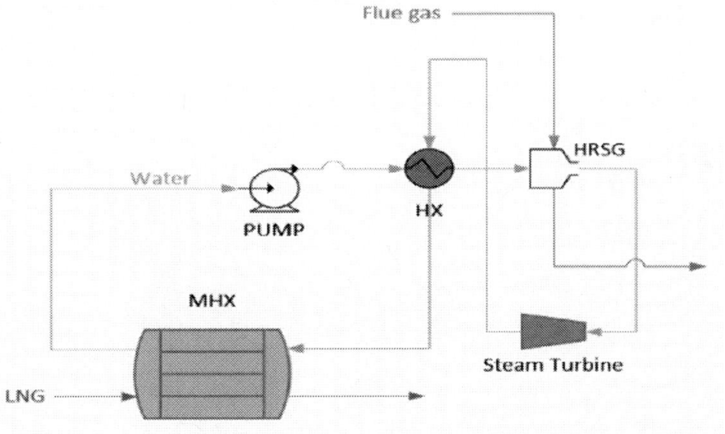

Fig.4 Flowsheet of the ST

By calculating the exergy efficiency and energy efficiency, the system's efficiency can be accurately evaluated. The exergy efficiency of the GT is 57.85%, and that of the HRSG is 71.46%. If only the work done by the GT is considered, the energy efficiency is 16.52%. Including the work from the ST increases the energy efficiency to 21.96%, an improvement of 5.44%.

The sensitivity analysis aims to study the relationship between the number of trays in the HPC and LPC and the concentration of the oxygen distillate from the LPC to find the optimal number of trays. It also examines the relationship between the pre-cooling temperature of the gas, the ratio of air entering the HPC, the reflux ratio of the HPC, and the concentration of the oxygen distillate from the LPC to determine the optimal design of the CASU.

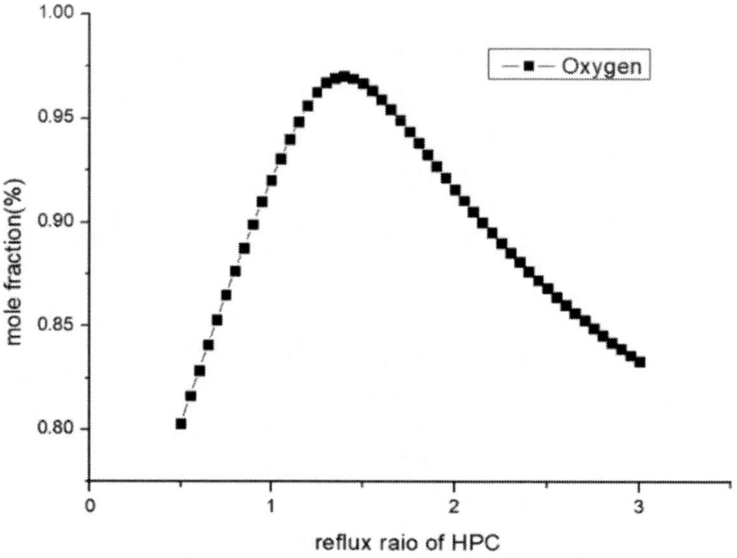

The results show that the net power generation of the plant is 63.03MW. The CASU requires 49.23MW of power, accounting for 78.11% of the plant's power generation, with 42.58MW of cooling load saved by utilizing the cold energy from LNG regasification. The carbon capture process requires 4.12MW of power, accounting for 6.53% of the total power generation, with 34.47MW of cooling load saved by using the cold energy from LNG regasification. This demonstrates the significant advantages of utilizing LNG regasification cold energy in units requiring substantial cooling.

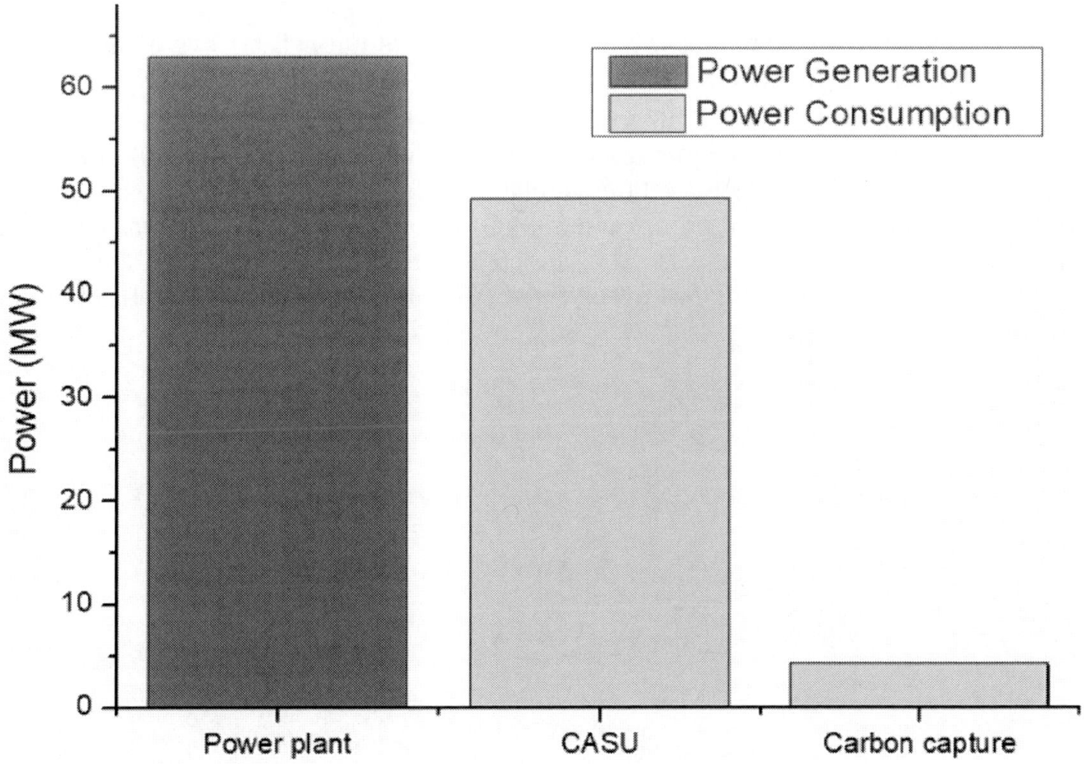

Keywords: LNG regasification; Cold energy; Oxy-fuel combustion; Carbon capture.

379f Fast Pyrolysis of Raw and Acid-Washed Cattle Manure in a Fluidized Bed Reactor for the Production of Bio-Oil

Jae-Rak Jeon, Syarif Hidayat, Jinsoo Kim, Hyun-Tae Hwang, Seung-Soo Kim

Livestock manure is currently a waste biomass and has difficulties in treatment worldwide. In this study, raw cattle manure (CM) was pretreated with a HNO_3 solution to reduce ash content, and the effect of acid-washing pretreatment and particle size of CM and operating temperature on fast pyrolysis characteristics was investigated. The experiment was carried out using a fluidized bed reactor with a size of 0.25–1.70 mm at 450–550 °C for 1 h. Additionally, 100 g of silica sand was used as the bed material, and the nitrogen flow rate was maintained at $3.0xU_{mf}$ (U_{mf} = 4.5 L/min). At 500 °C, the highest bio-oil yield of 31.18 wt% was observed from raw CM with a particle size of 0.71–1.19 mm. As the reaction temperature increased from 450 to 500 °C, the bio-oil yield increased, but a further increase to 550 °C decreased the yield due to the secondary decomposition of bio-oil. Under the same conditions (0.71–1.19 mm and 500 °C) where the highest yield was obtained from raw CM, the bio-oil yield from acid-washed CM increased to 42.15 wt%. Finally, the compositions of the gas and bio-oil obtained through fast pyrolysis of CM were systematically characterized using GC, GC-MS, ^{13}C-NMR and TGA.

379g Integrated Production of Methanol and Hydrogen: A Decarbonization Framework for Petrochemical Industry

Usama Ahmed

Methanol and hydrogen, regarded as cleaner fuel alternatives, present significant potential for reducing carbon emissions in the petrochemical industry. This research aims to establish a framework for process integration to concurrently produce methanol and hydrogen from vacuum residue while minimizing sulfur and carbon emissions. Two distinct process models were developed for this investigation. In the first scenario, vacuum residue undergoes gasification using oxygen and steam, resulting in syngas that is processed to yield both methanol and hydrogen. The second scenario follows a similar process model but places greater emphasis on methanol production from vacuum residue. Both models underwent comprehensive techno-economic comparison, considering factors such as methanol and hydrogen production rates, specific energy requirements, carbon conversion, CO2 emissions, overall process efficiencies, and project feasibility. The comparative analysis indicates that the second scenario, which prioritizes methanol production, offers several advantages. It reduces specific energy requirements by 86.01% compared to the first scenario and decreases CO2 emissions by 69.76%. Overall, the second scenario demonstrates superior project feasibility, exhibiting enhanced process performance and reduced production costs compared to the first scenario.

379h Synthesis of Surfactants and Evaluation of Their Potential for Enhanced Oil Recovery Applications

Masooma Nazar, Muhammad Israr, Syed Muhammad Shakil Hussain, Muhammad Shahzad Kamal, Mohammad Alotaibi, Mohanad Fahmi

Enhanced oil recovery (EOR) methods hold considerable promise for extracting residual oil after primary and secondary extraction processes. These methods commonly involve the application of various chemicals, including surfactants, alkalis, nanofluids, polymers, and co-solvents, to modify intermolecular interactions at the rock-oil-water interface. By changing wettability, reducing interfacial tension (IFT), and altering viscosity, these chemicals facilitate improved oil mobility. However, conventional EOR chemicals often face stability issues and substantial retention challenges.

This study aims to evaluate the effectiveness of recently developed surfactants as EOR agents, particularly under challenging reservoir conditions characterized by high temperatures and salinity levels. The surfactants were synthesized using a simple method, involving the amidation and quaternization. The successful synthesis of surfactants was confirmed through NMR and FTIR analysis.

These surfactants demonstrated notable solubility in both formation and seawater without experiencing phase separation. Thermal Gravimetric Analysis (TGA) data revealed decomposition temperatures exceeding 250°C, surpassing typical reservoir temperatures. The interfacial/surface studies revealed similar and/or better performance of locally synthesized surfactants compared to commercial substances.

The exceptional properties exhibited by these surfactants suggest significant potential for application in high-temperature, high-salinity reservoirs.

379i Reaction Behaviors during Induction Period of Crox for Fluidized-Bed Propane Dehydrogenation

Dae Sung Park, Hawon Park, Yong-Ki Park

Although a fluidized bed dehydrogenation process with a short residence time is desirable for saving energy, it is challenging to develop a catalyst with high activity and propylene selectivity. Chromium oxide has been identified as a suitable catalyst for propane dehydrogenation (PDH). However, it exhibits an induction period during which COx is formed before the formation of propylene within the very short time-on-stream (a few seconds). In this study, we developed a reaction system for monitoring products on the second timescale and investigated the processes involved in the induction period of PDH, and distinguished the different types of oxygen species derived from chromium oxide. We could establish the correlation between the initial selectivity to propylene and the type of chromium oxide species in the chromium catalyst supported on alumina by O_2-TPO, pulse-reaction, and Raman analyses. Non-selective oxygen species that cause deep oxidation of propane to COx during PDH were generated when the density of chromium on the support was high. The study indicated that most of the propane gas was converted to CO_2 or CO because of the accumulated lattice oxygen over the monolayer on the dispersed chromium species in the catalyst, while chromate species that were well dispersed on the alumina support contributed to high selectivity to propylene in a few seconds of the reaction. We also demonstrate that appropriate reduction of the catalytic CrO_x species can decrease the amount of non-selective oxygen species and that the calcination of the catalyst in an inert atmosphere can enhance product selectivity. In addition, based on this study, propane dehydrogenation was implemented in a bench-scale riser-type fluidized bed reactor using a chromium catalyst.

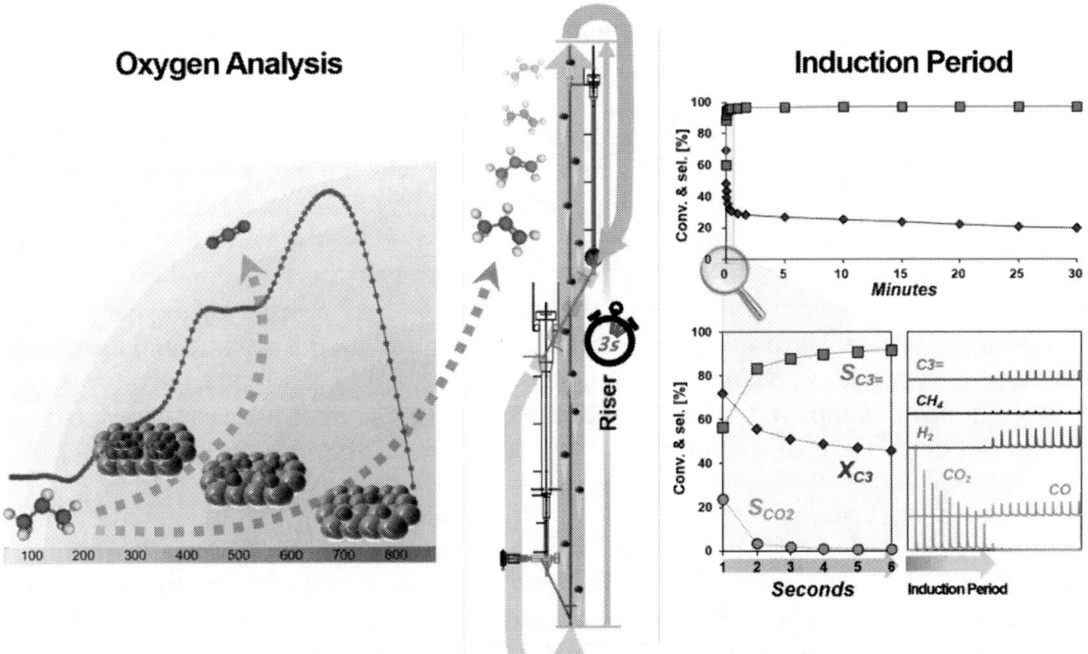

Fast-fluidized Propane Dehydrogenation

379j Hydrogenation of Pyrolysis Fuel Oil over Pt/MCM-41 Catalysts

Jong-Ki Jeon, Seung Kyo Oh, Sooyeon Hong, Yehee Kim, Seyeon Won

Steam cracking of naphtha is one of the most important petrochemical processes during the production of light olefins such as ethylene, propylene, butene, and butadiene. Pyrolysis fuel oil (PFO) is a heavy residue obtained from naphtha steam cracking plants. It is generally used in petrochemical plants as a process fuel. It is also noteworthy that the number of carbon atoms (C_8-C_{15}) in PFO is similar to that of aviation fuel. Nevertheless, the high aromatic content in PFO reduces the fuel quality and leads to the formation of unwanted emissions in the exhaust gas. The hydrogenation of PFO is an important process to upgrade its properties and make it more suitable for use as a component of aviation fuel.

Recently, platinum catalysts have been widely adapted for use in the hydrogenation of PFO due to their high activity and good selectivity in this reaction. These catalysts are designed to increase the active surface area and promote efficient mass transfers, leading to improved reaction rates and better product selectivity. This work aims to investigate the effect of Pt/MCM-41 catalysts on the hydrogenation of the polycyclic aromatic hydrocarbons included in PFO for the production of aviation fuel.

MCM-41 was synthesized using cetyltrimethylammonium bromide (CTABr) and sodium silicate solution. After dissolving 12.15 g of CTABr in 140 g of distilled water, the 50 g of sodium silicate solution (20 wt.% SiO_2) was added. The reaction mixture was heated to 100 °C in an oven for 24 h. After adjusting the pH to 10 with 50% acetic acid solution, the mixture was reacted again at 100 °C for 24 h. The product was washed with HCl/EtOH mixture, dried at 100 °C and finally calcined at 550 °C for 4 h.

The bead-type Pt/MCM-41 catalysts were prepared using two methods. In the first method, a mixture of MCM-41 powder, the organic binder and inorganic binder was initially formed into a bead (2 mm in diameter) through bead maker. Then, chloroplatinum hydrate (Aldrich) was impregnated into the MCM-41 bead using the incipient wetness method. The prepared bead-type catalyst was calcined at 400 °C for four hours, and it was termed 'Pt/MCM-41(AI) catalyst'. In the second method, chloroplatinum hydrate (Aldrich) was impregnated on the MCM-41 powder by incipient wetness method. Subsequently, catalyst forming was carried out after adding the organic and inorganic binders to prepare the bead-type catalyst (2 mm in diameter). The prepared bead-type catalyst was calcined at 400 °C for four hours to complete the catalysis of the bead-type Pt/MCM-41(BI) catalyst. The Pt loading amount of the of catalysts is 1.0 wt%.

Nitrogen adsorption analysis was performed using BELL SORP-mini II manufactured by BEL Japan. A 0.1 g sample was loaded into the cell, and after pretreatment at 200 °C for 8 h, adsorption-desorption isotherm was obtained at -196 °C. The specific surface area of the sample was calculated using the BET method. To measure the amount of acid cites and acid strength of the catalyst, ammonia-temperature programmed desorption (NH_3-TPD) analysis was performed using a BEL-CAT-B device (BEL Japan). Temperature programmed reduction of hydrogen (H_2-TPR) was performed using a BEL-CAT-B device from BEL JAPAN. Platinum dispersion was determined by CO-chemisorption experiments using Autochem II 2920 device (Micromeritics).

Crude PFO was received from Yeochun NCC Co., Ltd (Rep. of Korea). This crude PFO needs to distil out because it contains heavier polyaromatic hydrocarbons, which can lead to coke generation and catalyst deactivation during subsequent reaction. A total amount of 50 wt% PFO (PFO-cut) was obtained through distillation at 170 °C under vacuum of 20 mbar, which was used as a feedstock for the hydrogenation reaction. The majority of PFO-cut consists of aromatic hydrocarbons, with alkanes and cycloalkanes scarcely found. Naphthalene and its derivatives are presents as a main component (49.7%) along with alkylated benzene (19.1%), indene and its derivatives (16.3%), biphenyl (5.6%), fluorene (1.9%), and phenanthrene (1.3%). When the components are classified based on the number of rings, the fraction of bicyclic aromatic hydrocarbons is most prevalent at 72.5%, while the fractions of monocyclic aromatic hydrocarbons and tricyclic aromatic hydrocarbons account for 19.1% and 3.5%, respectively.

The hydrogenation of PFO-cut was carried out in a trickle-bed continuous-flow reactor. The reactor was manufactured using a SUS tube with an inner diameter of 1.3 cm and a length of 10 cm, which was used as a trickle-bed reactor. A 1 mm glass bead was filled in the upper and lower parts of the reactor, and 5 cc of the catalyst was filled between the glass beads and loaded into the reactor. The catalyst was reduced at 400°C under hydrogen flow. PFO-cut was injected using a high-pressure metering pump, and the hydrogen flow rate was controlled using a mass flow controller. The reactor pressure was constantly regulated through a back pressure regulator.

Liquid product samples were taken at regular intervals, and analyzed by gas chromatography (ACME 6100 GC, Young In Chromas) equipped with HP-5 column (30 m × 0.32 mm × 0.25 μm) and FID. In addition, a qualitative analysis of liquid sample was carried out using a gas chromatography-mass spectrometer detector (GC-MSD, Agilent 5937, Agilent Technologies, Inc.) equipped with an HP-5 column (30 m × 0.32 mm × 0.25 μm).

The yield of saturated cyclic compounds in PFO-cut hydrogenation did not decrease considerably over Pt-MCM-41 catalyst until ten hours of time-on-stream had elapsed, indicating that catalyst deactivation did not occur. As the time-on-stream increased from two to ten hours, the change in the carbon number distribution was negligible.

When the reaction temperature was increased from 200 °C to 300 °C, the yield of the saturated cyclic hydrocarbons increased monotonically, but when the temperature was raised further, the hydrogenation of PFO-cut instead decreased. Because the hydrogenation reaction of aromatic compounds is an exothermic reaction, the higher the temperature becomes, the less thermodynamically favorable the reaction becomes, and the equilibrium conversion rate is lowered. As a result, the optimal reaction temperature to maximize the conversion of PFO-cut under the given reaction conditions was determined to be 300 °C.

The Pt/MCM-41(BI) catalyst consistently maintained an average yield of about 37% of saturated cyclic compounds during the ten-hour hydrogenation reaction, representing higher yield than that of the Pt/MCM-41(AI) catalyst. This is attributed to the fact that the Pt/MCM-41(BI) catalyst has larger pore volume and pore size than those of Pt/MCM-41(AI) catalyst. In addition, the Pt/MCM-41(BI) catalyst has a larger number of acidic sites and an better reduction ability at low temperatures compared to the Pt/MCM-41 powder catalyst, resulting in an increased yield of aviation fuel.

The carbon number distribution of PFO-cut, the reaction raw material, was distributed between C_{10} and C_{15}. It was found that a cracking reaction occurred during the hydrogenation reaction, with a shift toward light hydrocarbons. The most abundant compound among the PFO-cut hydrogenation reaction products was a cycloalkane with a two-ring structure, and C_8-C_{14} hydrocarbons were included as the main products. The cyclic saturated hydrocarbon fraction with 12 carbon atoms was the most abundant in the PFO-cut hydrogenation product; this form is known to be suitable as the main fraction of aviation fuel.

379k Catalytic Oxidation Desulfurization of Dibenzothiophene By Bronsted-Lewis Acid Ionic Liquids

Ran Liu, Yali Yao, Xinying Liu, Fatang Li

Introduction: Sulfur oxides formed by burning S-compounds in transportation fuel led to serious pollution to the environment. [1] Ionic liquids used in the desulfurization of fuel performs excellent desulfurization capability and has many advantages, such as mild reaction conditions, simple operation, and a high desulfurization rate of multi-sulfide. [2-3] Four kinds Bronsted-Lewis ionic liquids with different metal chloride ($[HO_3SC_3NEt_3]AlCl_4$, $[HO_3SC_3NEt_3]ZnCl_3$, $[HO_3SC_3NEt_3]FeCl_4$ and $[HO_3SC_3NEt_3]CuCl_3$) were investigated as catalyst in the oxidative desulfurization.

Materials and Methods: 1,3-propanesultone, HCl, triethylamine, acetone, anhydrous $ZnCl_2$, anhydrous $FeCl_3$, anhydrous $AlCl_3$, anhydrous $CuCl_2$, H_2O_2, n-octane, dibenzothiophene, glacial acetic acid, isopropanol, 1,4-benzoquinone, 4-amino-2,2,6,6-tetramethylpiperidine and 5,5-dimethyl-1-pyrroline N-oxide. For the desulfurization experiment, 10 mL model oil or actual diesel mixed with certain amount of ILs, H_2O_2 and AcOH were added into a three-necked flask under vigorous stirring for reacting a certain time, and using oil bath to adjust the different reaction temperature. During the reaction, the sulfur content in upper phase was periodically separated and detected by sulfur analyzer. The sulfur content was measured by THA-2000S Ultraviolet Fluorescence Sulfur Analyzer. And the sulfur types were detected by GC-FPD.

Results and Discussion: The effect of DBT removal by four ILs was investigated. The results showed that the catalytic activity of the four ILs followed the order of $[HO_3SC_3NEt_3]AlCl_4 > [HO_3SC_3NEt_3]ZnCl_3 >> [HO_3SC_3NEt_3]FeCl_4 > [HO_3SC_3NEt_3]CuCl_3$. In the two-phase oxidation desulfurization system, a very small amount of $[HO_3SC_3NEt_3]AlCl_4$ showed a remarkable catalytic ability for the oxidation of dibenzothiophene (10 mL model oil with 1000 ppm sulphur content, 1mL H_2O_2, 2mL acetic acid, 55°C, 20min, 0.25g IL. According the analysis of the potential ROSs by quenching experiments and Electron Paramagnetic Resonance (EPR). There are three types of ROSs, $^1O_2$, •OH, and $•O_2^-$, in the system were detected. The desulfurization system can be recycled 4 times without noticeable decrease in activity (Table 1). $[HO_3SC_3NEt_3]AlCl_4$ showed excellent recycling performance. This IL-ODS system was applied to actual diesel and achieved a good desulfurization effect (Table 2). In summary, Among these four ILs, $[HO_3SC_3NEt_3]AlCl_4$ shows a remarkable catalytic activity. The EPR detection and quenching experiments verify the $^1O_2$ is in charge of the oxidation of DBT. IL is not only to promote the generation of ROSs in the reaction system, but also to act as a reaction medium between the oxidants and sulfides. The process and mechanism for catalytic oxidation of DBT in IL-ODS system was proposed in Fig. 1.

Table 1. Recycling of IL on removal of DBT in model oil

Cycle	DBT removal (%)	Cycle	DBT removal (%)
1	100	3	87.5
2	96.8	4	77.9

Table 2. Desulfurization effect of actual diesel

Entry	Sulfur content (ppm)	DBT removal (%)	Yield of oil (%)
Actual diesel oil	726	--	--
Only by extraction	658	9.3	>99
IL-ODS system	146	79.8	>99

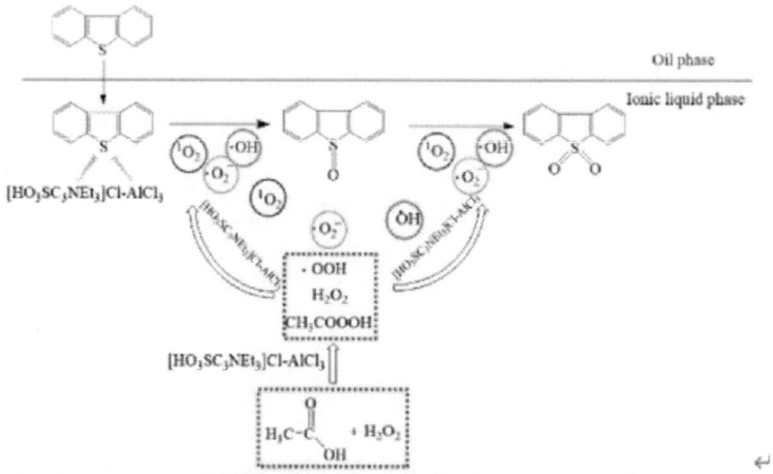

Fig. 1. The proposed process of DBT catalytic oxidation in IL-H_2O_2-AcOH reaction system.

3791 Novel Technological Roadmap from Fossils to Clean and Cheap Fuels Manufactured Domestically and Distribute

David Judbarovski

Our Technological roadmap from fossil fuels to clean and cheap fuels, manufactured domestically and distribute.

Abstract

Novel technological roadmap from carbon fossil fuels to fuels and their different applications, can be domestically and distribute manufactured (oil, gas and greatly recycled graphite) from carbon fuels, and water vapor, and carbon dioxide, and can be used too to produce solar photovoltaic cells and free hydrogen, for energy supply of any designed schedule, the all being cheap and clean and practically abundant and inexhausted.

Background

Oil and gas and coals now reaching dangerous level to environment, but known by their attractive ratios: energy to volume, energy to weight, energy to cost, and energy to price, even if cartel exporters increase market ratios: prices to cost up to in order-of-magnitude. All those initiate folly and dreams about hydrogen or ammonium as global panacea, and 'smart grids' for power supply, and about huge solar and wind electricity to be integrated in national grids. All those problems can be solved by our novel technological roadmap from fossil fuels to clean and cheap fuels manufactured domestically and distribute for distributed consumers.

Claims

Claim 1. Fischer-Tropsch process (Germany, a middle of 1920-th) to produce motor's fuels from carbon based fuels (e.g. coals, lignite or others ones) and the water vapor, $C(s) + H2O(g) = CO + H2$, named by 'syngas', to produce the oil, $CO + 2H2 = CH2 + H2O$, is combined with Boudouard reaction (France, 1905) converting carbon monoxide into carbon dioxide + graphite, $CO = 0.5 \times (CO2 + C)$ at quite mild conditions, and totally balanced in $1.5\ C + H2O = CH2 + 0.5\ CO2$, manufacturing the oil, is named by me as Judbarovski process (Israel, 2023), while carbon based fuels combustion provide heat energy, and here CH2 is abbreviation of the oil.

Claim 2. The oil manufactured by Claim 1 is decomposed thermally into hydrogen and more clean carbon (graphite) to recycle the dirty C (carbon) of Judbarovski process of Claim 1 and greatly cheapening it, simultaneously to eliminate new sites exploration, heavy mining, and long transportation for delivery, and to produce absolute free hydrogen, because expenses of it are relayed painlessly to the said graphite.

Claim 3. Carbon dioxide as a byproduct of Judbarovski process and free hydrogen produced correspondently to Claim 2 are raw materials for gas manufacturing by Sabatier process - $CO2 + 4\ H2 = CH4 + 2\ H2O$.

Claim 4, is the Claim 3, herein CO2 can be extracted from the air, when CO2 diffuses from the air and then diffused much more slowly through the water surface of water pond/ponds, and reacting with CaOH solution creating CaCO3' sediment on the said pond/ponds bottom and extracted for thermal decomposition in CO_2 and in CaO, going to be desolved in the water and return back in the said pond/ponds.

Claim 5. Oil and gas of Clam 1 and Claim 3 and hydrogen and carbon (graphite) of Claim 2 can be green and abundant and inexhausted very attractive raw material for chemistry. For example, ammonia produced by classical Haber-Bosh process.

Claim 6. Clean and cheap carbon of Claim 2 is used to produce solar photovoltaic cells.

Claim 7. Cheap and clean, and practically abundant and inexhausted fuels manufactured following to Claim 1, Claim 2, Claim 3, Claim 4 separately or in their different combinations can be used to support and help renewables as heat as electricity to satisfy any designed schedule of energy supply being distributed for distributed consumers, to satisfy if war times and terrorist attack or so.

379m Development of a Calculator for Gibbs Energy of Formation

Isaac Gamwo

Oil and gas production and geothermal wells can easily develop inorganic scales making flow assurance a critical issue for their production. We have extended predictive scale formation models to high-temperature, low-pressure (HTLP) conditions to proactively mitigate deposition of minerals in HTLP pressure conditions. Here, we have developed an interactive user interface and sub-routines from a Mathematica code to support a user-friendly web calculator for aqueous thermodynamic properties.

This presentation will display the QRcode that users can scan then select the species of interest, the temperature and pressure to calculate the Gibbs free energy values needed to predict the mineral scale formation. The user-friendly calculator will be available on the NETL's Energy Data Exchange (EDX) to provide easy access to these recent advances for the scientific community.

379n Sustainable Solution for the Effective Removal of Poly Aromatic Hydrocarbons from Petroleum Fractions.

Zainab Alaithan, Mohammed Alkhunaizi, Emad Shafi, Nadrah Alawani

This study investigates the development and application of a porous carbon as a sustainable material for the removal of poly aromatic hydrocarbons (PAH) from heavy petroleum fractions. PAHs pose an operational challenge in refining due to their toxicity and high molecular weight. While advanced refining techniques such as hydrocracking and hydroisomerization reduce PAH levels, complete elimination is not always feasible, leading to environmental contamination and compromised products quality. Recognizing the prevalence of PAHs in crude oil and the associated risks, our research focuses on developing sustainable and effective materials for removing PAHs from heavy oil fractions. This study optimizes the structure and surface properties of activated porous carbons to enhance PAHs absorption. Experimental results demonstrate that our activated porous carbon significantly lowers PAH content in heavy petroleum streams, meeting rigorous industry standards and contributing to environmental safety. Our finding suggests that our porous carbon offers a sustainable and effective solution for reducing PAH concentrations in petroleum derived products. This innovation not only marks a significant advancement in refining technology but also provides a path toward producing cleaner and safer petroleum products.

379p Electrohydrodynamic Processing in Biosolvents Production from Waste Polylactide

Sabina Wilkanowicz, Peng-Kai Kao

We present the design, creation, and development of immobilizing calcium oxide (CaO) catalyst onto polymeric mats for environmentally friendly biosolvents production. The polyacrylonitrile (PAN) fiber mat was prepared by electrospinning technique, in which immobilized CaO particles were physically entrapped on the fibers. Immobilized and free-flowing CaO catalysts were later successfully used in the transesterification of waste polylactide into lactic acid methyl esters (MeLa). Entrapped CaO particles in the PAN fibers were able to be re-used up to 5 times in the conversion of waste polyester into biosolvent,. The use of electrospinning technique allows the creation of an easily separable and reusable catalyst, which provides a sustainable method for future transesterification reactions of waste polylactide.

Moreover, produced in proposed method ester (MeLa) was successfully used as a solvent in ecologically friendly demulsifier - an additive commonly used in the petroleum industry, especially in protected marine areas. By changing the traditionally used solvents, like isopropanol or butyl glycol, to MeLa we were able to produce ecologically friendly additive with excellent oil-water demulsifying properties. Furthermore, produced additive was safe for microorganisms isolated from marine and sweet waters and presented inhibition properties on microbial corrosion.

267a Modeling Vapor Liquid Equilibrium in Novel Binary Mixtures Using Machine Learning

Mohammad Alam, Charles McGill

Industries rely heavily on accurate equations to predict the behavior of gas and liquid mixtures for separation processes. However, due to reliance on specific data for regression, existing equations such as NRTL cannot produce models on novel mixtures. This limitation has prompted the development of alternative approaches such as UNIFAC, a group additivity model which uses functional group compositions to predict properties without requiring previous data of a specific mixture. UNIFAC does require pairwise interaction parameters to produce good models, but these parameters may not exist for uncommon pairs. UNIFAC is hence constrained by its reliance on group additivity and the rigid functional group categories, which don't allow nearby functional groups to influence each other.

In this study, we provide an alternative to UNIFAC by leveraging machine learning models on known VLE databases. Our approach aims to offer a more flexible representation compared to traditional and group additivity models, allowing projection to novel mixture pairs and considering neighboring functional groups in the calculation. We apply different model structures, utilizing the functional forms of traditional models as the model predictive output, thereby maintaining thermodynamic consistency of predictions and ensuring a smooth prediction output.

Furthermore, we incorporate simultaneous prediction of vapor pressure in the model, enabling predictions for species outside the range of available Antoine's coefficients. We benchmark our trained model against UNIFAC and other traditional models trained on the same data. The model is then evaluated by testing the predicted location of azeotropes against known locations. Our results demonstrate a greater ability to generalize novel mixtures and condition extensions of existing mixtures. We also include uncertainty quantification into the model to estimate the bounds of applicability for model prediction.

267b Improving the Thermodynamic Consistency between Vapor Pressure, Heat of Vaporization, and Liquid and Ideal Gas Heat Capacity Predictions for Associating Chemicals

Cassandra Guffey, Neil Giles, Thomas Knotts IV, Wade Vincent Wilding

Accurate values for thermophysical properties are vital to process design and optimization. Many researchers have used the simultaneous correlation of vapor pressure, heat of vaporization, ideal gas heat capacity, liquid heat capacity, and liquid and vapor volumes as a powerful method to model these properties with higher accuracy and thermodynamic consistency. However, achieving such accuracy becomes notably challenging for chemicals that associate. Choices in correlation equations, equations of state and their parameters, and quantum and statistical mechanics methods used in calculation of ideal gas heat capacity significantly impact accuracy. In this work, we tested several equations of state and vapor pressure correlation equations to determine optimal methods for simultaneous prediction of the aforementioned properties, specifically focusing on associating chemicals. By addressing complexities with associating chemicals and offering improved methodologies for simultaneous correlation, this research contributes to improving the reliability of thermophysical property predictions for associating chemicals.

267c Assessing the Accuracy of Predicting the Cloud Point of Biodiesel Using Ideal Solution Theory

Anthony Malshyti, Michael Senra

Biodiesel is a promising replacement for petroleum based diesel in a number of different applications. Although biodiesel offers numerous advantages over petrodiesel, one of its major disadvantages is its inferior cold flow properties. Previous work has shown that the challenge with the cold flow properites is strongly correlated to the composition of the biodiesel, which primarily contains fatty acid methyl esters (FAME). Taking advantage of this fact, researchers have used a wide variety of thermodynamic models of varying complexities to conduct solid/liquid equilibrium analysis to estimate the cloud point of the biodiesel. The major issues that can cause variations in these models can include such parameters as degree of saturation, presence of non-FAMEs and co-crystallizaton. This work comprehensively looks at cloud point data from a wide variety of literature sources and analyzes them using the simplest model, the ideal solution theory. Work has shown that with a few exceptions, ideal solution theory predicts the cloud point quite accurately, particularly when taking into account the variability of measurements such as cloud point and the heat of crystallization.

267d A Novel Method for Correlating and Extrapolating Liquid Viscosity from Near the Melting Point to Near the Critical Point Using a Model Based on a Cotangent Function.

Brayton Young, Neil Giles, Thomas Knotts IV, W. Vincent Wilding

Liquid viscosity is an important thermophysical property in both scientific and industrial applications. This study proposes a new method of extrapolation using a model based on a cotangent function. The approach to this new method focuses on the qualitative aspects of liquid viscosity correlations as a function of temperature. This method can work with a broad range of chemicals and families. Extrapolation requires limited data near the melting point of most compounds and proves adequate to temperature regions near the critical point. The simplicity of the method makes it ideal. Where other extrapolation methods rely on experimental data near the boiling point or complicated formulas that rely on empirical data, this method simply uses 3-4 regressed parameters and molecular weight. The method can also be used to determine the reliability of experimental data when comparing similar compounds or compounds within the same family. The versatility and applicability of this method make it a valuable tool when extrapolating or predicting liquid viscosity.

267e Increasing the Yield of High-Value, Liquid-Crystalline-Forming Pyrene Pitch Oligomers

Jack Schmidt, Mark Thies

Carbonaceous pitch, an underutilized byproduct of fossil-fuel processing, is comprised primarily of polycyclic aromatic hydrocarbons (PAHs). These high-carbon materials are of great interest in the aerospace, automotive, and construction industries. Pitch itself can be derived from both coal tar and petroleum processing; however, both pitches are difficult to characterize and thus most effectively utilize. The molecular composition of fossil-fuel-derived pitch is broad and consists of hundreds of different compounds. Nevertheless, pitches are still useful for the synthesis of advanced carbon materials. Their char yields have been shown to have a direct correlation to many of the mechanical properties of the pitch-based advanced carbon materials. Previous workers (including the recently deceased pioneer giant of synthetic pitches, Isao Mochida) have oligomerized naphthalene and anthracene to create model pitches. Although these compounds, and in particular naphthalene pitches, have excellent properties for materials application, they fragment easily during synthesis and thus also have broad molecular weight distributions.

Our research has focused primarily on pitches derived from pyrene, a relatively new model pitch that recent work from Chen et al. 2020 shows as an excellent analogy to petroleum pitch systems. Pyrene addresses many of the problems from other model pitches because it essentially does not fragment; thus, it has a narrow molecular weight distribution. However, controlling the degree of oligomerization of pyrene pitches is of significant interest because of its impact on the mesophase content (i.e., liquid crystallinity) of these pitches. This work explores utilizing a new reagent (to be presented) for increasing the degree of oligomerization of pyrene-derived pitches. In this study, reaction conditions are held constant to allow direct comparison between pitch derived via this new reagent and that derived via the existing technique. We have observed that pyrene pitches synthesized from this new reagent have a softening point and char yield over twice that of standard pyrene-derived pitch, with similar increases in composition increases of the liquid-crystalline trimer.

We also explored the effects of the reaction conditions (reaction time, pressure, and temperature) on the kinetics of the Scholl reaction – and thus the properties of the derived pitch. By using a combination of supercritical extraction (SCE) and gel permeation chromatography (GPC), oligomeric compositions of each of the synthesized pyrene pitches are being obtained. Another encouraging result is that the new synthesis technique requires less than a quarter of the time to produce the desired product. In summary, our new reagent allows us to synthesize a pyrene pitch that is a more useful model for the development of advanced carbon materials where softening point, char yield, and liquid crystallinity are all of primary importance.

267f Prediction of the Density and Viscosity of Visbroken Heavy Oils Using Regular Solutions and Expanded Fluid Models.

Jose Angel Beleno Alcazar, Camilo Andres Lopez Lago, Amirabbas Abbaspour, Florian F. Schoeggl, Harvey W. Yarranton

Visbreaking is a mild thermal cracking process that significantly reduces oil viscosity and moderately reduces oil density. The higher the severity (temperature and residence time), the greater the viscosity reduction. However, the practical extent of viscosity reduction is limited by the final product stability; that is, its tendency to precipitate asphaltenes or form coke. Hence, predictions of product properties and stability are required to design and optimize visbreaking processes. The oil can be characterized in different pseudo-components. The most common strategies include the use of boiling points or solubility classes. The composition of the oil can later be used in different models to predict their density and viscosity.

Regular solution and Expanded Fluids are an example of models that can be used to predict the density and viscosity of heavy oils. The regular solution mixing (or volumetric) rule assumes that the intermolecular forces between like and unlike molecules are the same; hence, there is no volume change and or enthalpy change upon mixing. The mixture density is then the volumetric average of the component densities. Expanded Fluid (EF) model correlates viscosity to density. The fluid density is determined from the dilute gas viscosity and a departure function for the liquid viscosity. The departure function is similar to free volume theory except that relationship of liquid viscosity to inverse density is similar to a double exponential function instead of an exponential function. This correlation traces continuously the full range of single-phase viscosity in the gas, liquid, and critical or supercritical regions.

Recently, the density and viscosity of visbroken products from a Western Canadian bitumen were matched with correlations based on a DSARA characterization using Regular Solutions and EF. DSARA (distillates, saturate, aromatic, resins and asphaltenes) separates the oil into the distillable fraction (boiling point < 367 °C AET) and the SARA components of the residue. An advantage of this approach is that the volatile components are captured with the distillate fraction and the residue has a low volatility, providing a more consistent SARA assay. In addition, the two fractions that exhibit the largest changes with visbreaking (distillates and asphaltenes) are represented directly in the characterization. A significant limitation of this approach was that the DSARA composition of the visbroken products were necessary to be measured. Lately, a model to correlate DSARA composition to visbreaking conversion was developed.

The purpose of this study was to evaluate the Regular Solutions and EF models to predict the density and viscosity of visbroken oils using only the conversion and initial feed composition as input. Nine different feeds including a Colombian heavy crude, three Western Canadian bitumens, two partially deasphalted bitumens, a fully deasphalted bitumen, and a vacuum residue fraction were thermally cracked in a continuous visbreaker at constant pressure over a range of conversions. The Simulated Distillation assay, gas yield, pentane-insoluble asphaltene content, toluene insoluble (TI) content, density, viscosity, distillate content, and residue SARA composition of each feed and product were measured. Conversions from 5 to 46% were explored. The gas yield was below 1.5 wt% and the TI content was negligible in all cases. The density and viscosity of the visbroken products were predicted with Regular Solutions and EF fluids model and compared to the experimental values. The average absolute deviation for the density was below to 2.5 kg/m³ and the viscosity average relative deviation was below to 30%.

Phase Behavior Modeling of Gasoline Systems

Xiangwei Liu, Independent Consultant, Vancouver, BC

Extended Abstract

The vapor pressure of a gasoline is an important physical property that affects the performance of the fuel when it is used in internal combustion engines. A number of vapor pressure test methods have been standardized by the American Society for Testing and Materials (ASTM) to quantify the volatility characteristics of the automotive fuels. The prospect of using rigorous thermodynamic principles in conjunction with physical property models to represent the standard test methods and to assess the volatility characteristics of automotive gasoline is discussed. Mathematical models involving various bubble point, dew point, and flash calculations have been developed. In particular, the prediction of the Reid vapor pressure and the representation of the differential vaporization behavior of gasoline mixtures are illustrated. Furthermore, a comparison of the performances of different equations of state in representing the phase behavior of mixtures of selected hydrocarbons and oxygenated additives is presented. Specifically, the UNIFAC model has been incorporated into the Peng-Robinson equation of state with the van der Waals mixing rules to predict the phase behavior of hydrocarbon mixtures with ethanol or methanol.

267h Prediction of Cetane Number Using Molecular Graph Modularity and Functional Groups in Artificial Neural Networks

Cristopher Arvizu, Jorge Aburto, Elias Martinez-Hernandez

PREDICTION OF CETANE NUMBER USING MOLECULAR GRAPH MODULARITY AND FUNCTIONAL GROUPS IN ARTIFICIAL NEURAL NETWORKS

Cristopher Arvizu Cano[1], Jorge Aburto[2], Elias Martinez-Hernandez[2]

[1] Faculty of Chemistry, National Autonomous University of Mexico

[2] Energy Efficiency and Biofuels Division, Mexican Institute of Petroleum

The cetane number is a measure of the ignition quality of a diesel fuel in a compression ignition engine and it is experimentally determined with an engine test standard (ASTM D613). A fuel with a high cetane number has a short ignition delay period and causes fast release and combustion of fuel after being injected into an engine where the air has been compressed and heated to a high temperature. A molecule is good candidate as an additive or fuel for diesel engines if it has a high number cetane. In this work, we developed a model to predict cetane number to identify candidate molecules derived from biomass by using molecular graph modularity and functional group counts as inputs to train an artificial neural network (ANN). This approach has been demonstrated as promising in physical property predictions of properties such as viscosity. A database from literature was collected and processed with the final data set containing 519 compounds of different class (alcohols, aldehydes, alkenes, aromatics, carboxylic acids, cyclo-alkanes, esters, ethers, furans, iso-alkanes, ketones, lactones, n-alkanes and polyfunctional), which were split into training and test sets. The ANNs were trained in Matlab using the ANN Toolbox with a single hidden layer and varying the number of hidden neurons to minimize the root mean squared error (RMSE). To test the benefit of including modularity, models consisting of only functional group counts were developed and compared. In addition, the prediction was compared against other existing models in the literature. After model reduction by removing functional groups with lowest relevance, the best model was the one incorporating modularity with a RMSE=5.61. Results of the RMSE showed that the new models decreased the prediction error by up to 45% or maintain the same level but with the advantage of not requiring other features such as calculated quantum chemical descriptors. The parity plots for each compound classes showed a correlation value >=0.9. In general, we found that molecules with higher modularity values and presence of long carbon chains, linear ester and ether groups lead to higher cetane number. Finally, we applied the model to screen potential derivatives from biomass finding candidates for centane enhancing additives.

477a Microwave Catalytic Decomposition of Natural Gas to Clean Hydrogen and Carbon Nanomaterials Via Modular Process Intensified Production

Changle Jiang, Jianli Hu, Brandon Robinson, Juan A. Lopez-Ruiz, Yuan Jiang, Robert A. Dagle, Sherafghan Iftikhar, Luke Neal, Fanxing Li, Xinyuan Zhu, Wenyuan Li, Xingbo Liu, George Skoptsov

Microwave-driven catalytic decomposition of natural gas has a great potential to be adopted for the hydrogen production industry. The process took advantage of the convenience of microwave heating, which provided lower catalyst bulk temperature, instant start and stop on reactions, as well as faster heating and cooling. The process generated only hydrogen as a gas product, and solid carbon as by-product. The gas separation became much easier because all it took was to separate the hydrogen from the unreacted methane. Regarding solid products, carbon nanotubes were the new pursuit for the targeted product. It was of high value, which has huge potential to be used in both carbon fiber production as well as carbon-based semiconductors. The United States Department of Energy (DOE) has set up a goal of reducing the hydrogen selling price to \$1/kg in a decade. This team started a DOE funded project to work toward this goal from 2020. As of today, the team has developed two series of catalysts for the microwave-driven catalytic decomposition of natural gas. One series of catalysts was to use carbon nanotube as support, Ni as active catalytic site, and Fe, Cu, or Pd as a promoter. The conversion of methane under this series of catalysts ranged from 30% to 55%. These catalysts were excellent microwave absorbers. The promoters helped extend the activity of Ni catalyst to 6~10 h when compared to Ni only catalyst lasting about 1 h. The other series of catalysts was SrTiO supported Ni. The conversion of methane using this catalyst is comparable to the previous catalyst. However, the microwave attenuation of these catalysts still needs to be improved because the reduction process of these catalyst played a significant role in microwave absorption ability. Lastly, MgO was introduced to this SrTiO to help disperse the Ni nanoparticles, which has improved the performance of the Ni-SrTiO system. The improved SrTiO catalyst achieved methane conversion ranging from 20% to 50%. An alternative route for improving the performance of Ni-SrTiO was under development by another team member. La and Ca were used to replace Sr, resulting in a LaCaNiTiO catalyst, which has achieved a methane conversion of around 15%. Carbon nanotubes/nanofibers of 10-80 nm diameter were produced as solid carbon. Our team members from Pacific Northwest National lab have optimized the catalyst loading as well as conducted high resolution transmission electronic microscope for imaging the carbon products. In addition, they have simulated the hydrogen selling price as around \$2/kg using our technology, which could be as low as \$1/kg if base-growth of carbon nanotubes could be developed. We have our commercial partner working on the pilot testing of the commercial Ni based catalyst for scaling up the process. The effluent gas from microwave plasma conversion of methane was used as feedstock. The effluent underwent catalytic decomposition to generate hydrogen and carbon nanotube at a larger scale. Currently, a pilot reactor of 36 kW has been installed. The testing is in process. The eventual goal of this project is to achieve microwave only driven reaction for catalytic conversion of methane to hydrogen, remove the carbon emissions, and significantly reduce the cost of clean hydrogen production.

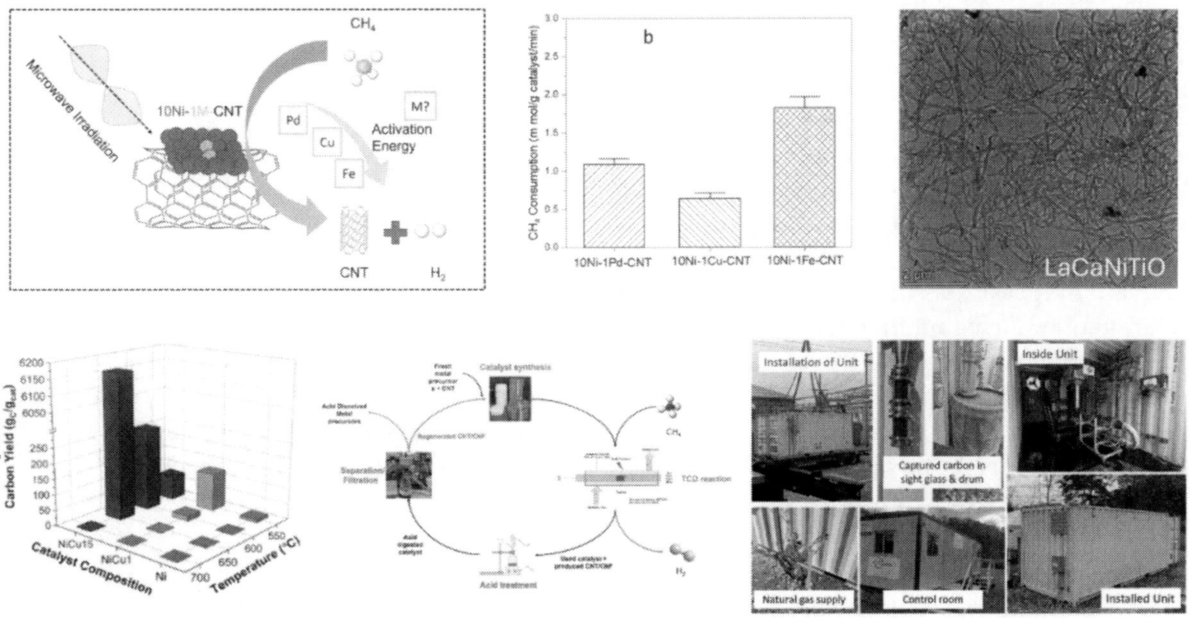

Microwave Irradiation
10Ni-1M-CNT
CH$_4$
Pd
Cu
Fe
M?
Activation Energy
CNT
H$_2$
b
CH$_4$ Consumption (m mol/g catalyst/min)
10Ni-1Pd-CNT
10Ni-1Cu-CNT
10Ni-1Fe-CNT
LaCaNiTiO
Projected Carbon Yield
Catalyst Composition
Temperature (°C)
NiCu15
NiCu1
Ni
Catalyst synthesis
Separation/ Filtration
TCD reaction
Acid treatment
Installation of Unit
Inside Unit
Captured carbon in sight glass & drum
Natural gas supply
Control room
Installed Unit

477b Effect of Support on Iron-Based Catalyst Toward CO_2-Free H_2 Production from Methane Pyrolysis

Divakar Reddy Aireddy, Daniel Haynes

The global demand for low-emission hydrogen production is increasing rapidly due to the potential for hydrogen to substitute conventional fossil fuels in energy-related applications. Among practical technologies such as steam reforming of methane (SMR), water electrolysis, and methane pyrolysis, methane pyrolysis has the most potential to produce economical- and near-zero-emission hydrogen. The reaction during methane pyrolysis produces a solid carbon by-product that can be monetized to offset production costs. However, this route is not without research challenges, and the selective formation of high-value carbon remains a critical issue. Supported nickel, iron, and cobalt, have been widely investigated for methane pyrolysis. Although iron-based catalysts are less active than Ni, iron catalysts offer less resistance to coke formation, are inexpensive, and have a longer lifetime.

One area not sufficiently explored is the effect of support materials other than Al_2O_3 for the iron-based catalysts. Therefore, we investigated a detailed variation of support materials to study their influence on the catalytic performance. Moreover, this study offers a fair and direct comparison among all catalytic materials which were synthesized and studied under identical conditions. This direct comparison of the obtained results is important because results reported in different publications are often not comparable due to very different reaction conditions. Combining these with the in-situ characterization will advance the understanding of catalytic carbon formation for the production of CO_2-free H_2 production from methane pyrolysis.

477d Induction-Heated Molten Salt Reactor for Methane Pyrolysis

Zhennan Ru, Chenghao Wan, Calvin Lin, Connor Cremers, Pinak Mohapatra, Dorothy Mantle, Ariana Hofelmann, Juan Rivas, Jonathan Fan

As the world shifts away from fossil sources as a chemical feedstock and fuel, much attention has been paid towards clean energy infrastructure based on the hydrogen economy. Hydrogen stands as one of the most promising green fuel and chemical substitutes for several reasons: 1) hydrogen has a high heat of combustion; 2) upon combustion, the primary byproduct is water and no carbon dioxide is produced; 3) hydrogen has a high energy density; and 4) hydrogen is an essential feedstock for many chemicals such as ammonia and plastics. Amongst the several clean energy concepts for hydrogen production that are being explored, methane pyrolysis has become a focal point for study because it produces hydrogen from a carbonaceous feedstock and the solid carbon byproduct is easy to sequester. Methane pyrolysis is also economically attractive, as it is a scalable thermochemical process and the solid carbon byproduct can potentially be utilized to further enhance its economics.

Methane pyrolysis is a challenging process to implement at scale because it is a highly endothermic reaction and requires exceptionally high temperatures on the order of 1000 °C and above. In addition, methane pyrolysis poses challenges when performed in the presence of solid catalyst materials, as carbon generated from the pyrolysis process can coat the catalyst surface and cause deactivation. Along these lines, molten salt reactors for methane pyrolysis have emerged to tackle this very issue. Carbon generated from pyrolysis in a molten salt reactor either dissolves into the liquid or stays at the gas-liquid interface, which prevents catalyst deactivation. The solid carbon can be retrieved after removing residual salt, and since the salts used are water soluble, the residual salt contamination can be removed by flushing, which improves carbon product quality. There is nonetheless the challenge of heating molten salt pyrolysis reactors using green energy sources, such as renewable electricity, to very high temperatures.

We present an inductively heated molten salt reactor that can heat chloride-based molten salts to temperatures above 1000 °C. In our reactor concept, alternative magnetic fields generated using a helical magnetic coil directly couple to the molten salts themselves, producing volumetric eddy currents within the molten salts that dissipate as heat. The concept takes advantage of the fact that molten salts are ionic liquids and are electrically conductive, and they can therefore serve as a volumetric heating susceptor. Our concept in wireless molten salt heating has multiple advantages compared to alternative concepts. First, it circumvents the need for immersion heaters, which run into challenges pertaining to chemical and temperature compatibility with the caustic, high temperature molten salt bath. Second, it can be tailored to supply heat to the molten salt bath in a volumetric fashion, eliminating heat transfer bottlenecks in the methane pyrolysis reaction. Third, it is scalable and can readily adapt to meter-scale reactors and megawatt power levels. Fourth, it is relatively safe to operate.

To experimentally realize our inductively heated molten salt reactor concept, we implement a 9 cm-diameter reactor system containing a 1:1 sodium chloride and potassium chloride mixture as the salt bed. Heating is performed using a combination of resistively heated wires and magnetic induction coils surrounding a molten salt reactor. The resistive heating jacket, made from kanthal wires, is used to initially heat the solid salt to a molten state, and it is required because solid salts are not electrically conductive. Once the salts reach a molten state, the resistive heater is turned off and the 23 cm-diameter, 30 cm-long helical copper induction coil is activated and continues to deliver power to the salt. The heater and coil are carefully designed to minimize parasitic induction heating of the resistive heater.

Co-design of the power electronics with the molten salt bed is essential to ensuring that the induction heating process is volumetric and efficient. Along these lines, we specify the induction frequency to be 2.5 MHz, from which the skin depth (i.e., length scale specifying magnetic field penetration into the molten salt) is approximately the reactor radius. AC impedance measurements of the molten salts are used to quantify the AC resistivity of the molten salts at different temperatures and confirm that our skin depth condition is maintained. Coupling efficiency, which is defined as power delivered to the molten salt divided by total power delivered to the molten salt, resistive heater and helical coil, is measured using large signal analysis to be 92% and is consistent with theoretical calculations. Finite element multiphysics simulation tools coupled with experimentally measured temperature profiling confirm that the inductively heated molten salt bed supports a volumetrically uniform heating profile.

477e Economic Comparison of Methane Pyrolysis Processes for Hydrogen Production Based on Various Molten-Metal Catalysts

Semie Kim, Young-Il Lim, Uen-Do Lee, Youn-Bae Kang

Interests in clean energy sources are increasing to simultaneously pursue carbon neutrality and energy supply. Hydrogen, a clean energy source, does not emit greenhouse gases such as carbon dioxide during combustion and generates only water and heat. Hydrogen is categorized into gray, blue, green, and turquoise hydrogen, depending on the production technology and carbon dioxide emissions. Turquoise H_2 technology, which produces hydrogen and solid carbon from methane, has the advantage of less energy to produce hydrogen compared to water electrolysis and less CO_2 emissions. The methane pyrolysis without catalyst requires endothermical energy above 1300 °C. Solid metal-based, carbon-based, and molten-metal (MM) based catalysts have been used to reduce the reaction temperature. The buble column reactors (BCRs) with MM catalysts offer high mass and heat transfer efficiency. Various MM catalysts were reported to decrease the operating temperature and increase the methane conversion.

In this study, a one-dimensional (1D) MMBCR model was used to calculate the CH_4 conversion in the BCRs with MM catalysts such as tin, gallium, copper, Ni_{27}-Bi_{73}, tellurium for CH_4 pyrolysis. The process flow diagrams (PFDs) with carbon separation were presented for a hydrogen production capacity of 200 kt/yr. Economic indicators such as hydrogen production cost, total capital investment (TCI), and total production cost (TPC) were compared for the H_2 production processes using different molten-metals. The effects of the methane conversion and economic values on MM catalysts were analyzed.

477f Hydrogen from Natural Gas with No Greenhouse Gas Emissions

Eric D. Wachsman

Membrane reactor technology holds the promise to circumvent thermodynamic equilibrium limitations by *in-situ* removal of product species, resulting in improved chemical yields. Recent advances in mixed-conducting oxide-membrane technology present the possibility for a dramatic reduction in the cost and environmental impact of converting petroleum, coal and biomass derived feed stocks to hydrogen and other "value added" hydrocarbons.

We have developed single-step gas to liquid reactors that convert natural gas to C_{2+} products with high yields and no unwanted oxidation byproducts. Conversion of CH_4 to simultaneously produce H_2 on one side of a mixed protonic-electronic conducting $SrCe_{0.7}Zr_{0.2}Eu_{0.1}O_{3-\delta}$ membrane and C_{2+} hydrocarbons (C_2 and aromatics) on the other using a direct non-oxidative methane conversion (DNMC) iron/silica catalyst is demonstrated. Further, by adjusting the operating conditions (e.g., temperature) we can tailor the hydrocarbon products between C_2's (e.g., C_2H_4) and aromatics (e.g., C_6H_6).

Recently we focused our membrane reactor development on hydrogen production with negligible greenhouse gas emissions, e.g., for the $2CH_4 \rightarrow 2H_2 + C_2H_4$ reaction all the carbon is tied up in the ethylene co-product. If electricity from today's grid (0.4 kg CO_2/kWh$_e$) is used to provide necessary enthalpy this reaction can produce hydrogen at only 0.06 kg CO_2/kg H_2 which is orders of magnitude smaller than either SMR or electrolysis on today's grid. Moreover, if we take the carbon footprint of ethylene production into consideration this process for hydrogen production is actually carbon negative.

477g Palladium Membrane Technology for Low-Carbon Hydrogen Production

Alaa Albasry, Hassan Khalaf, Mohamed Draze, Abdulbari Alqarni, Minseok Bae, Stephen Paglieri, Aadesh Harale

Hydrogen is expected to play a critical role in the transition to a sustainable energy system. It can help in promoting energy security and decarbonizing different sectors, especially those that are hard to abate such as long-haul transport and heavy industry. However, the majority of H_2 is currently produced through conventional processes, which are responsible for significant carbon emissions. Therefore, technologies for low-emission H_2 production are needed to unlock H_2 full potential as a sustainable energy carrier.

Herein, we present an overview of the palladium membrane technology and its development at Saudi Aramco's Research and Development Center as a pre-combustion capture technology and promising route for low-carbon H_2 production. In this technology, the reforming reactor is designed to house H_2 perm-selective palladium-based membranes, thus facilitating simultaneous H_2 production and separation in one unit. Enhanced conversions beyond the thermodynamic limitations are achieved with high H_2 purity (>99%), all while operating at reduced temperatures of 500-600 °C compared to > 900 °C in the conventional reforming processes. Moreover, the produced CO_2 stream creates a single emission point of concentrated and pressurized CO_2, which can be captured more efficiently at reduced costs and energy penalties.

477h Life Cycle Assessment of Methane Pyrolysis through a 3-Stage Catalytic Membrane System

Yuming Wen, Iskandar Halim, Jie Chang, Sibudjing Kawi, Chi-Hwa Wang

A new methane catalytic pyrolysis system was proposed using a 3-stage catalytic membrane system to produce hydrogen and carbon nanotubes (CNTs) from natural gas. In the designed process, the methane catalytic decomposition was conducted at a temperature of 550 °C using a nickel catalyst, which is significantly lower than the steam reforming temperature range of 700 to 1000 °C. Moreover, pressure swing adsorption (PSA) is avoided by applying a membrane to separate the hydrogen from the flue gas. The produced CNTs can act as carbon materials, such as activated carbon. Based on this, a series of studies were conducted to investigate the environmental friendliness of the proposed system. Aspen Plus V14 was first used to simulate the proposed system, and the results were used to conduct a cradle-to-grave life cycle assessment (LCA) to compare the environmental impacts of the 3-stage catalytic membrane system and the commercial steam reforming process. The LCA was conducted using SimaPro 9.5 with the ReCiPe 2016 Midpoint (H) method. The results showed that the global warming potential of producing 1 kg of hydrogen through the 3-stage catalytic membrane system and steam reforming in Singapore was 9.46 and 11.7 kg CO2-eq, respectively. Based on the results, it is of interest to further optimize the process experimentally and build a demonstration plant to investigate the feasibility of the proposed system.

597a Sustainable Chemicals Program Leveraging Stranded and Underutilized Natural Gas As Feedstocks

Daniel Haynes, Jared Ciferno

Sustainable Chemicals Program Leveraging Stranded and Underutilized Natural Gas as Feedstocks

Daniel Haynes[1], Jared Ciferno[2]

[1]National Energy Technology Lab, 3610 Collins Ferry Rd. Morgantown, WV 26505

[2]U.S. Department of Energy, 1000 Independence Ave., Washington D.C. 20585

In the pursuit of addressing the climate crisis, the Department of Energy has aligned its initiatives with the Nations aggressive emissions targets needed to reach Net-Zero emissions by 2050. Included in this are the bold action strategies that support the pillars of decarbonization which include the development of transformational technologies that can lead to game changing innovations and their demonstrations to ensure market readiness and adoption. The viability of these new technologies rests on their ability to provide reductions in decarbonization through improved energy efficiencies, electrification, sustainable feedstocks and energy sources, as well as CO2 capture, sequestration, and storage.

In this light the Office of Fossil Energy and Carbon Managements Methane Mitigation Division supports the broader DOE initiatives by addressing methane emissions from the hydrocarbon fuel sector, with efforts to improve emissions detection, quantification and mitigation capabilities over the production, transport, and storage these fuels.

One of the most significant sources of hydrocarbon emissions is associated gas flaring. To exemplify the extent of this problem, between 2011 and 2021 almost 3.5 trillion cubic feed of natural gas was vented or flared across the United States alone and produced 356 million metric tons of CO2 [1]. While there are many reasons for flaring, it is primarily done for economic reasons due to limited pipeline takeaway capacity in the remote areas the wells are located (e.g., Permian and Bakken regions). Recognizing the environmental and economic benefits, the Division of Methane Mitigation Technologies has identified these wasted resources as an opportunity to be converted into industrially relevant value-added chemicals and fuels.

The United States chemical sector produces roughly 274 million metric tons of CO_2 per year and is expected to grow as this is considered one of the hardest sectors to decarbonize because of wide assortment of processes and operations [2]. The Division of Methane Mitigation sees the ability to reduce the chemical sectors emissions by converting these underutilized natural gas resources into high-value, high-demand, and low carbon foot-print chemicals such as ammonia, ethylene, propylene, and methanol. Our analysis suggests that advanced conversion technologies could lead to a reduction in national green house gas emissions from ammonia production by 48%, ethylene production by 12.5%, or from methanol production by 100% respectively (assuming all underutilized gas is being used to produce either of these chemicals).

The Division of Methane Mitigation Technologies is seeking to expand its Natural Gas Conversion Program into a Stranded and Underutilized Natural Gas Conversion More Sustainable Chemicals sub-

Program. This program will lead research efforts to develop modular technical solutions to convert waste underutilized gas into sustainable chemicals that offer advanced net-zero pathways. The program will have major themes that include waste feedstock utilization, innovative carbon-efficient chemical conversion processes, and demonstration of advanced processes with a cross-cutting theme of understanding sustainability implications. The themes of this program support align with DOEs Clean Fuels and Products Shot™ to meet the goal of 50% carbon-based chemicals3.

References:

[1] IEA, 'Direct CO2 combustion emissions from flaring and flaring intensity in the Net Zero Scenario, 2010-2030,' International Energy Agency, 10 July 2023. [Online]. Available: https://www.iea.org/data-and-statistics/charts/direct-co2-combustion-emissions-from-flaring-and-flaring-intensity-in-the-net-zero-scenario-2010-2030. (Accessed 2024)

[2] Industrial Decarbonization Roadmap (energy.gov) (Accessed 2024)

[3] Clean Fuels & Products Shot™: Alternative Sources for Carbon-based Products | Department of Energy (Accessed 2024)

597c Microwave-Driven Non-Oxidative Methane Coupling Using Mo/CeO_2 catalyst for Ethylene Production: Insight into the Effect of Metal Promoters.

Brhanu Kelali Desta, Alazar Araia, Jianli Hu

With the growing need to transition to cleaner and more sustainable energy sources, the battle for efficient hydrocarbon conversion technologies has reached a critical point. Methane, a key component of natural gas (~95%) and shale gas (~70%), has become a focus for innovative solutions due to its abundance and environmental concerns (Xiao & Varma, 2018). Therefore, the strategic conversion of methane to ethylene, the key building block in the chemical industry, is critical to the long-term use of natural gas and shell gas utilization and the global move to net zero emission (Gao et al., 2019). Currently, ethylene is produced through the generation of syngas and methanol, which are then converted into ethylene. However, this approach is expensive and environmentally unfriendly, as it produces undesirable byproducts such as CO and CO_2, contributing to greenhouse gas emissions. As a result, direct conversion of CH_4 can potentially reduce the production complexity and minimize the environmental effects. Non-oxidative methane coupling (NOCM), which avoids traditional oxidative procedures, opens up new possibilities for developing CH4 conversion technologies to produce more sustainable and environmentally friendly products. In addition, it also has the potential to improve selectivity and energy efficiency.

Despite its economic and environmental benefits, direct, NOCM to value-added compounds has major thermodynamic limits. Therefore, designing an effective catalyst is crucial to resolving methane conversion difficulties. Furthermore, employing microwave irradiation to heat the catalyst is an effective technique to reduce energy consumption. A well-designed catalyst increases methane conversion efficiency by allowing selective chemical reactions, lowering unwanted byproducts, and increasing desired outcomes.
Microwave irradiation for catalyst heating is a concentrated and energy-efficient strategy. Unlike other heating methods, microwave irradiation selectively heats the catalyst, resulting in efficient energy transfer at the reaction sites. This application of microwave technology in NOCM not only improves methane conversion and ethylene selectivity, but it also solves environmental issues by enabling for operation on intermittent renewable energy sources. Such focused energy delivery can significantly reduce the overall energy input required for bond cleavage in methane coupling, addressing the high energy demand of strong carbon-hydrogen bonds.

In this study, we conducted NOCM using a Mo/CeO_2 catalyst under microwave irradiation at relatively moderate temperatures (700°C). Various metal promoters were introduced into the primary catalyst system, Mo/CeO_2, to assess their impact on the direct conversion of methane and ethylene production. The findings revealed that catalysts containing metal promoters exhibited higher ethylene selectivity compared to the unpromoted Mo/CeO_2. Among them, Cs-Mo/CeO_2 emerged as the most efficient catalyst for ethylene and C_2-selectivity, achieving rates of 60% and 90%, respectively, with a reasonable CH_4 conversion rate of approximately 22%. The catalyst's properties were studied using diverse physicochemical characterization methods, including CO chemisorption, temperature-programmed reduction (TPR), X-ray diffraction (XRD), and Brunauer–Emmett–Teller (BET) surface area analysis.

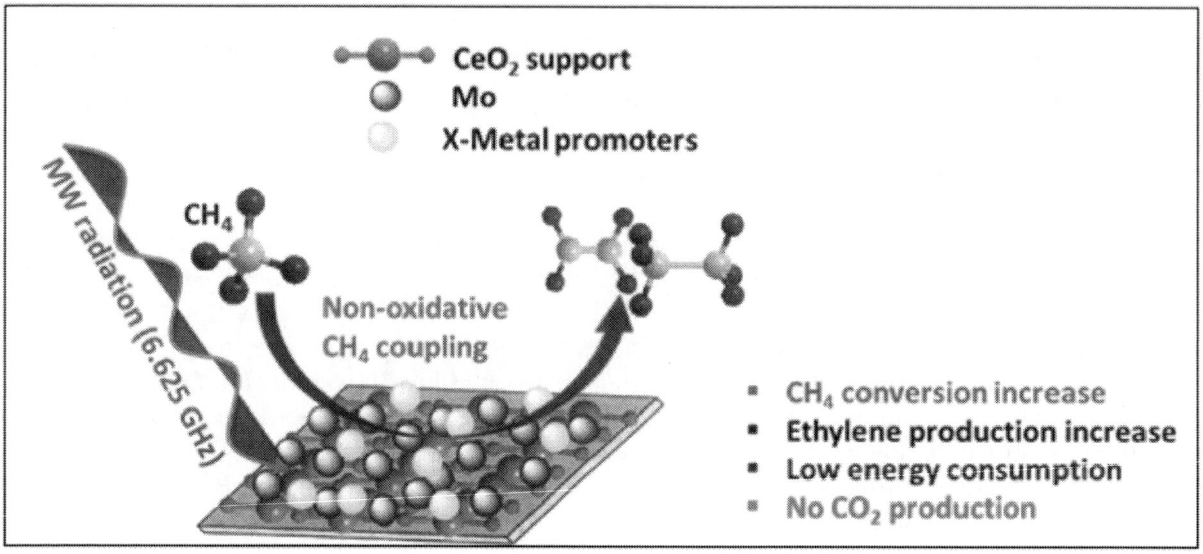
CeO$_2$ support
Mo
X-Metal promoters
MW radiation (6.625 GHz)
CH$_4$
Non-oxidative
CH$_4$ coupling
CH$_4$ conversion increase
Ethylene production increase
Low energy consumption
No CO$_2$ production

597b Conceptual Design of a Novel Industrial Complex for Simultaneous Production of Electricity, Urea, and LNG with Extremely Low Greenhouse Gas Emissions

Ying Liu, Song Wang, Qiang Xu, Yinlun Huang

Natural gas, as the fastest growing primary energy and raw material resource, has been widely used in the chemical/petrochemical, residential, power generation, commercial, and transportation sectors in recent decades. To satisfy the globally increasing demand of clean power and fertilizer productions with the growing concerns of both industrial and environmental sustainability, the use of natural gas in large-scale manufacturing systems should be further investigated. In this regard, a new conceptual design of an industrial complex, namely the A^3UL complex, has been developed, simulated, and evaluated in this paper. The complex optimally integrates energy, power, and materials among five subsystems, including an Allam-cycle power plant (ACPP), an air separation plant (ASP), an ammonia production plant (APP), a urea production plant (UPP), and a liquefied natural gas (LNG) production plant (LPP). The A^3UL complex takes natural gas and air as the main feedstock to simultaneously produce electricity, LNG, urea, plus valuable byproducts such as high-purity carbon dioxide, nitrogen, and steam. In the meantime, the complex completely avoids nitrogen oxides emission and significantly reduces carbon dioxide emission, as compared with traditional productions of the above products. The operating feasibility and economic attractiveness of the complex have been virtually validated through rigorous modeling and simulation. The simulation results show that the overall revenue of the A^3UL complex will reach at least $43 per Mcf of natural gas utilization, which is much higher than any stand-alone process for power generation, urea production, or LNG production. The developed conceptual industrial complex can support the development of solutions for monetizing natural gas resources effectively, efficiently, and environmentally friendly.

597d Non-Catalytic Direct Partial Oxidation of Methane to Methanol in a Wall-Coated Microreactor

Kelly Cohen, Justin Blanchard, Paul Rodriguez, Kevin Kelly, James Dorman, Kerry Dooley

This project uses an integrated mixer-reactor-heat exchanger microtube scalable module for the non-catalytic direct partial oxidation of methane in natural gas to methanol at elevated pressures (70-95 bar), reaction temperatures of 380-460C, reactant CH_4:air ratios of 0.5-4.4 and reaction zone residence times on the order of 1 min.

Currently, methane in remote locations is often flared in large quantities rather than transported or converted to the more easily transportable methanol. However, if this gaseous methane were converted to liquid methanol at the wellhead, it could be transported in ways other than pipeline, or used onsite. For example, methanol is added to pipeline gas at remote locations to prevent hydrate formation in pipelines, and is a component used in water purification packages. In the case of hydrate formation, the direct methanol to methanol (DMTM) process could save ~\$500,000/year for one remote wellhead installation and reduce its carbon footprint by 10^4 metric tons of CO_2.

Additionally, methane is currently converted to methanol commercially via an expensive, energy intensive two-step process. First, the methane is steam-reformed to synthesis gas (CO and H2), which then undergoes a high-pressure catalytic conversion to methanol.

The microtube reactor/heat exchanger system for this DMTM process consists of a methane/air mixing, preheating and reaction sections and a quenching section. In the quenching section, the heat transfer is from tubes to shell. The shell-side flow rate is controlled to attain the desired reaction and final (~250°C) process fluid temperatures. The entire reactor, tubes and shell, are made of Inconel 625 alloy, a high Ni-Cr steel (60 wt% Ni, 22 wt% Cr). To passivate the metal and reduce rates of total oxidation, the tubes are carbide-coated in a process based on the decomposition of propane at ~700°C. Several other coating schemes were tested along with several other potential tube metals (e.g., sulfiding instead of carbiding, and also metals C1010 mild steel, 304L, 904L, Nichrome and pure Ni) in coating experiments using coupons. Other coatings, including nitrided carbide layers, are begin explored. The reasons for the choice of I625 and the carbiding process will be discussed.

With this system, one can control the five key parameters (temperature, pressure, residence time, wall inertness and methane:air ratio) to selectively produce methanol at greater than 8% per-pass yield (product of conversion and selectivity), under optimal conditions. For this module, these conditions were 80 bar, ~420C outer wall temperature, a methane/air molar ratio of 2.9 and a residence time of 0.8 min. At lower methane/air molar ratios good yields were still possible, but at lower methanol selectivities. So, while methanol can be made selectively in the microtube reactor system, its selectivity can vary over a wide range depending upon certain reaction conditions, and whille the carbide layer is not stable under long-term, repeated use, it is currently stable for about two days of operation.

597f Efficient Conversion of Ethane and Propane at Room Temperature and Ambient Pressure

Wenxuan Liu, Qi Lu

The direct conversion of light alkanes into value-added commodities is of great economic importance. At present, high temperatures (200-600 °C) and/or elevated pressures (typically 5-50 bar) are typically needed to facilitate the reaction, which requires large-scale infrastructures to accommodate the tortuous energetic pathways, hindering their implementation on small-scale and distributed productions. In our study, we report an aqueous reaction system capable of selectively converting light alkanes into olefins and oxygenates at room temperature and ambient pressure using Cu as catalyst and oxygen as oxidant. In ethane and propane activation under ambient conditions, we achieve comparable results to those in literatures that are obtained at elevated temperatures. Our system can convert ethane to ethylene and acetic acid with a high selectivity up to 97%, and propane mostly to propene with a selectivity up to 94%. Based on catalytic experiments, spectroscopic insights and density functional theory calculations, we put forward mechanistic understandings in which C-H bond is activated by surface oxide species generated during oxidation process forming alkyl groups as key reaction intermediates. To further promote alkane activations limited by the low solubility of gaseous reactants, we introduce a microfluidic reaction system with a Cu microtube serving as both the catalyst and the microchannel reactor. This innovative setup exhibits a 27-fold increase in reactivity compared to previous batch reactors, while maintaining propylene selectivity above 92%. Additionally, by combining alkane activations with an electrochemical system utilizing Cu plates as both electrodes and catalysts, we achieve consistent production of ethylene and acetic acid from ethane activation over a 10-hour run, with sustained selectivity above 96% and a concentration of acetic acid reaching 19 mM. These findings represent remarkable progress in the electrochemical activation of light alkanes under mild conditions. Our *in situ* spectroscopic investigations, analysis of partial pressure dependencies, and computational results provide a deeper understanding of the competitive adsorption behavior of oxygen and alkanes on the Cu surface. These strategies hold great promise for pioneering decentralized alkane conversion processes, enabling on-site and long-term utilization of light alkanes at ambient conditions.

597g Effect of Titania Structure and Particle Size on Titania-Supported Iridium-Ruthenium Catalysts in the Oxidative Transformation of Methane

Helena Weaver, Li-Yin Hsiao

Selective oxidation of methane to higher value chemicals is a desired albeit challenging pathway to utilizing natural gas, as, in the presence of oxygen, avoiding overoxidation to the most thermodynamically stable products, i.e. carbon dioxide and water, is difficult at best. Rutile iridium dioxide with (110) surface facets is particularly promising due to its potential for activating methane below room temperature. However, the presence of gas phase oxygen tends to block active sites, requiring higher temperatures for catalytic turnover and making it more difficult to avoid overoxidation. Our group has evaluated the effects of titania structure (anatase, brookite and rutile) on the complete and partial oxidation as well as dry reforming of methane over titania-supported pure and mixed iridium-ruthenium oxides to gain insight into methane activation and reaction over these catalysts. Low loadings of active metals were utilized (0.5% by weight) to maximize metal-support interactions and to limit the cost of expensive metals. Titania-supported ruthenium dioxide catalysts have a distinctly different behavior in these reactions compared with iridium dioxide catalysts on the same support. Under partial oxidation conditions, the IrO_x/TiO_2 catalyst is more active at lower temperatures than the RuO_2/TiO_2 catalyst (Figure 1a), while under dry reforming conditions, the RuO_2/TiO_2 is slightly more active than IrO_x/TiO_2 (Figure 1b). Furthermore, the titania structure does not significantly influence the catalytic activity over IrO_x/TiO_2 catalysts under partial oxidation conditions (Figure 1a), i.e., under oxygen-limited methane oxidation. In contrast, the activity in the complete oxidation of methane (excess oxygen) over IrO_x/TiO_2 catalyst depends not only on the titania crystal structure (anatase, brookite and rutile), but also the titania particle size (Figure 1c). These results indicate a complex influence of titania support structure on the catalytic properties, but also reveal a potential for tuning the methane activation and reaction over IrO_x-RuO_x/TiO_2 catalysts.

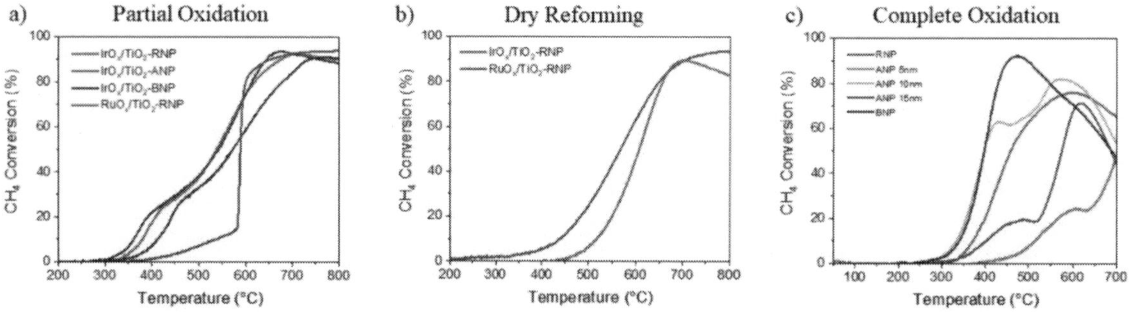

Figure 1. a) Partial oxidation of methane over IrO_x/TiO_2 and RuO_x/TiO_2 catalysts (CH_4:O_2 = 1:2). b) Dry reforming of methane over IrO_x/TiO_2 and RuO_x/TiO_2 catalysts (CH_4:CO_2 = 1:1), and c) complete oxidation over IrO_x/TiO_2 catalysts (CH_4:O_2 = 1:10). ANP = anatase, BNP = brookite, and RNP = rutile TiO_2 nanoparticles.

597h Regulating Lattice Oxygen Property of TiO_2 redox Catalyst for Enhanced Oxidative Dehydrogenation of Ethane

Xiaoqiao Zhang

Ethane oxidative dehydrogenation appears to be one of the most promising technologies in terms of potential industrial applications for ethylene production. However, creating a catalyst with improved selectivity without sacrificing catalytic activity remains a difficult issue. In this study, a systematic set of catalysts (V_5TiO_2, $Mn_7V_5TiO_2$, $Mn_2Ce_2V_7TiO_2$ and $Mn_4Ce_2Fe_6V_6TiO_2$) with redox performance were designed by machine learning. Different metals and their combinations modulated the lattice oxygen properties of titanium dioxide catalysts, thus achieving precise regulation of catalytic performance and product distribution. The results showed that $Mn_7V_5TiO_2$ had medium ethylene selectivity (55.5%) and high conversion rate (61.8%). The redox capacity was enhanced and both V^{5+} and the lattice oxygen species available to oxidative dehydrogenation reaction were enriched by incorporation of atomically dispersed Ce species. $Mn_4Ce_2V_7TiO_2$ was found to be highly selective for ethylene, with the selectivity of 94.4%. After optimizing the process conditions, $Mn_4Ce_2V_7TiO_2$ obtained higher ethane conversion (65.2%) and stable ethylene selectivity (89.6%). These results show that $Mn_4Ce_2V_7TiO_2$ catalyst has great application potential in the oxidative dehydrogenation of ethane.

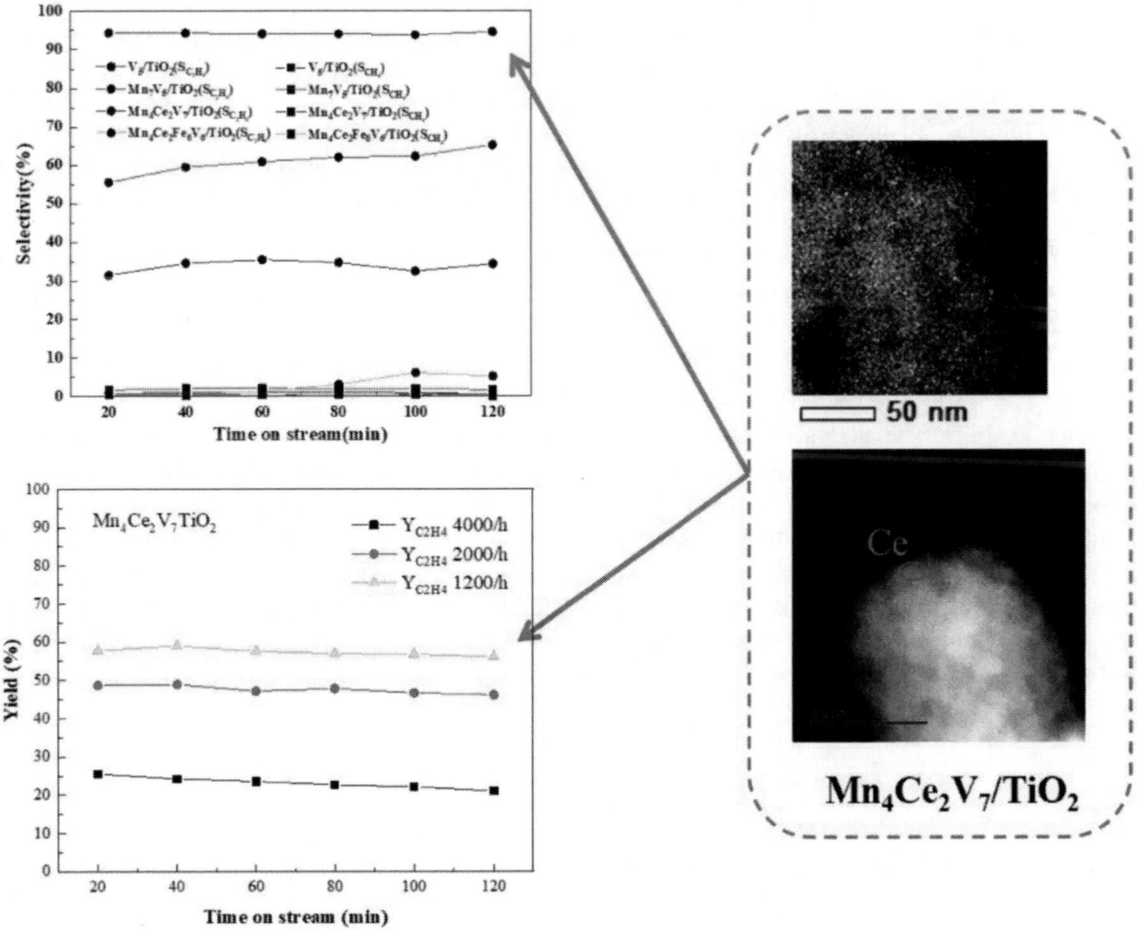

74a Experimental Study of Intrinsic Kinetics for the Pyrolysis of Polypropylene in a Microreactor Using Detailed Compositional and Principal Component Analysis

Barbara Alejandra Perez, Hilal Ezgi Toraman, Varaha Jonnalagedda

The U.S. produced 35.67 million tons of plastic waste in 2018, of which 3.09 million tons were recycled—while 26.97 million tons of plastic waste were landfilled, corresponding to 75% of all plastic waste produced in the U.S. in 2018. Polypropylene (PP) is widely used polyolefin due to its chemical and temperature resistance. The U.S. produced 8.15 million tons of PP waste in 2018, of which only 0.6% was recycled. Pyrolysis is a promising thermochemical recycling technique defined as the chemical transformation of a polymer by heating to a moderate temperature (400 – 600°C) under a non-reactive atmosphere in the absence or presence of catalysts. However, the successful implementation of pyrolysis on a commercial scale relies on having a mechanistic understanding to effectively manage changes in the feedstock characteristics without compromising the product quality and to design catalysts and processes that maximize the production of high-value added chemicals, fuels, and materials. In addition, detailed knowledge of the composition of the pyrolysis oils is fundamental to upgrading the pyrolysis oil in processes such as steam cracking for the production of monomers and other chemicals.

Studying the speciation data under experimentally controlled conditions that allow intrinsic kinetics is fundamental to developing (micro-)kinetic models for process development and optimization. In this work, we systematically studied the primary decomposition of PP under isothermal kinetically controlled conditions using a tandem microreactor coupled to a comprehensive two-dimensional gas chromatography with a time-of-flight mass spectrometer and flame ionization detector (Py-GC×GC-FID/TOF-MS). GC×GC is a powerful technique with superior resolution compared to traditional one-dimensional gas chromatography (GC). GC×GC allows a comprehensive identification and quantification of the pyrolysis products. Box-Behnken design of experiments (BBD) was used to evaluate the pyrolysis variables (i.e., pyrolysis temperature, sample size, helium flow rate, and particle size) along principal component analysis (PCA) to evaluate statistical differences between the product distribution of the experiments conducted. In addition, BBD border conditions and conditions beyond BBD were evaluated to identify the changes that extreme conditions cause to product distribution under enhanced heat and mass transport effects.

The analysis of 600 pyrolysis products identified and quantified in the pyrolysis of PP for the condition combinations evaluated in BBD resulted in statistical differences at temperatures of 460, 530, and 600 °C, with C_3 to C_{35} yields ranging from 39.88 2.08 to 66.61 3.26 wt.%. However, no statistical differences were found in the pyrolysis product distribution within the ranges of sample weight (50 – 150 µg), particle size (53 – 125, 125 – 300, and >300 µm), and carrier gas flow rate (100 – 300 ml min^{-1}) studied in the BBD experimental space. The analysis of the border conditions showed that a combination of the lowest carrier gas flow rate (100 ml min^{-1}) and the highest sample weight (150 µg) resulted in statistically different product distribution at the studied pyrolysis temperatures compared to the combination suggested by BBD. Conditions beyond the BBD, including a sample weight of 800 µg and a carrier gas flow rate of 30 ml min^{-1}, resulted in statistically different product distributions compared to the BBD. Notably, new products such as polycyclic aromatic hydrocarbons (PAHs) were identified. The findings from this work clearly show the importance of selecting appropriate pyrolysis parameters to obtain kinetically relevant experimental data, which in turn serves as a valuable guide for the development and validation of (micro-) kinetic models.

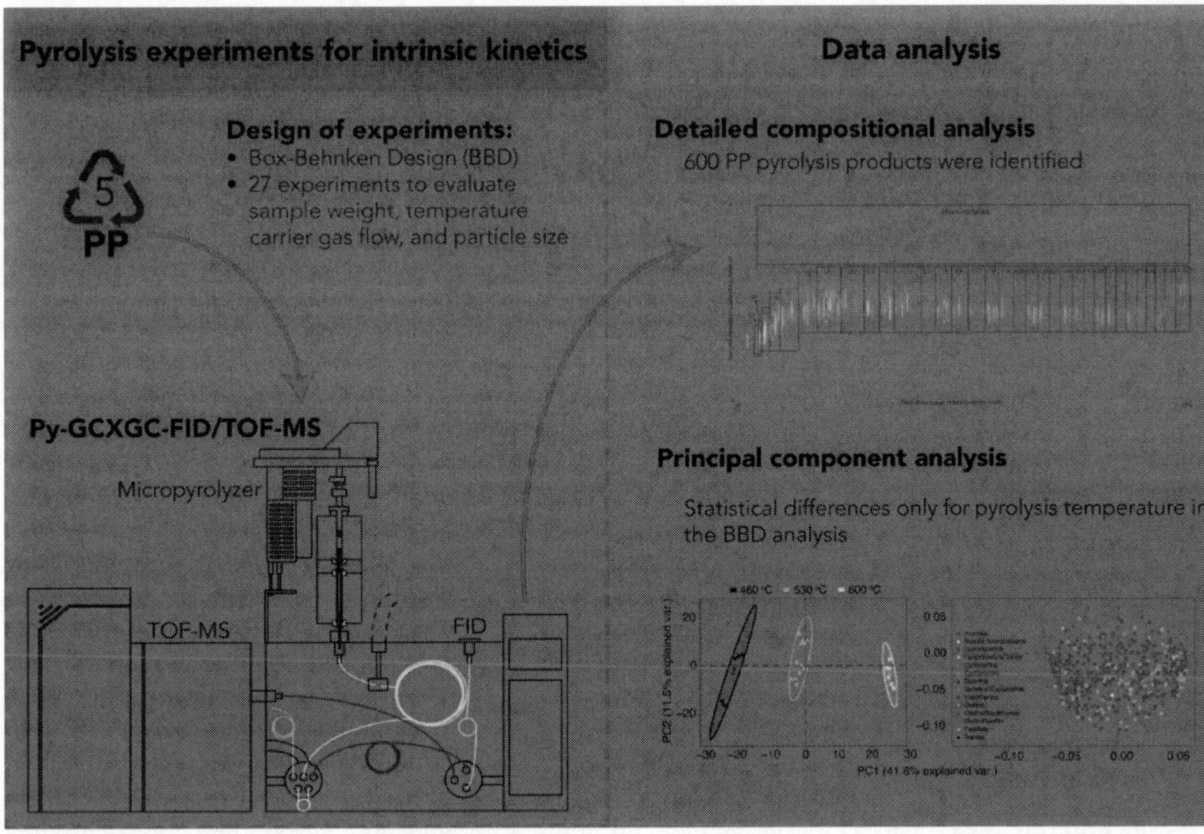
Pyrolysis experiments for intrinsic kinetics
Design of experiments:
• Box-Behnken Design (BBD)
• 27 experiments to evaluate sample weight, temperature carrier gas flow, and particle size
5
PP
Py-GCXGC-FID/TOF-MS
Micropyrolyzer
TOF-MS
FID
Data analysis
Detailed compositional analysis
600 PP pyrolysis products were identified
Principal component analysis
Statistical differences only for pyrolysis temperature in the BBD analysis
PC2 (11.5% explained var.)
PC1 (41.6% explained var.)

74b Strategies to Decarbonize Top 5 Industrial Chemicals and Its Implication on Greenhouse Gas Emissions

Arpit Bhatt, Ling Tao

The industrial sector is a key enabler for producing the chemicals that are intermediates to most commercial and household products used worldwide. Growing energy trends, increasing population, and global economic growth have led to supply chain imbalances that could potentially lead to increased uncertainty and future investment decisions. The petrochemical industry in U.S. is expected to increase in the next 5 years and to drive the production of building block chemicals. With the increasing energy demand and use of end products, the chemical industry has become increasingly vigilant about using alternative non-fossil feedstocks to meet supply chain demand. The production of chemicals from alternative sources like organic waste, carbon dioxide (CO2), plastics, and biomass offers a promising opportunity to reduce dependence on fossil fuels and improve the overall economic and environmental sustainability of the production process.

While previous studies have assessed a broad range of chemicals and their production from alternative feedstock sources to increase the deployment and scale-up the production of these chemicals, none of them provided recommendations on the prospective energy and carbon reductions that could result from the implementation of these alternate strategies. This knowledge gap on strategies to decarbonize the petrochemical industry and quantifying the greenhouse gas (GHG) emissions as compared to fossil-based routes can help lay out strategies to minimize carbon emissions and enhancing energy efficiency for industrial sector.

Our selection process considers market data, economics, strategies, and carbon and energy intensity to determine the top 5 chemicals (ethanol, propylene, benzene, toluene, and xylene). We also perform a comprehensive analysis to enable decision-making and help the industrial sector by providing strategies to decarbonize the chemical sector and align it with its sustainability goals, offering energy savings and reducing overall carbon footprints. Furthermore, we evaluate and compare the GHG emissions of alternative supply chain strategies in relation to traditional fossil-based methods. This analysis helps us understand the potential reduction in GHG emissions that can be achieved to meet the BETO target of cutting carbon emissions by 100 million metric tons (MMT) per year.

The results of our study indicate that it is possible to reduce carbon emissions in the chemical sector by employing alternative feedstocks and implementing a range of strategies. By considering factors such as feedstock availability and production capacity, an optimal solution can be achieved to meet the overall goal of reducing GHG emissions by 100 MMT/yr. The largest reduction in GHG emissions on a mass basis is associated with two processes: ethanol produced from lignocellulosic biomass via fermentation process and benzene-toluene-xylene mixture from wood chips via pyrolysis process. Due to the significant production capacities in the US, the two alternative processes result in a 91 MMT/yr reduction in GHG emissions compared to the baseline.

Recommendations for incorporating sustainable feedstock can be informed by analyzing biomass production routes, taking into account market dynamics, demand trends, regulations, and consumer preferences, to ensure commercial viability and relevance of proposed pathways. Quantifying energy and carbon savings allows for the demonstration of the tangible benefits of embracing more sustainable production practices.

74c Hydrogen Enhanced Conversion of Municipal Solid Waste to Methanol Using Natural Gas Pyrolysis: Potential Net GHG Reduction Impacts

Mohammad Ostadi, Daniel R. Cohn, Guiyan Zang, Leslie Bromberg

Avoiding methane emissions from landfilling of municipal waste and the need for low-carbon liquid fuels and chemicals are two important needs for decarbonization of methane emitting practices and the transportation and chemical industry sectors. Converting municipal solid waste (MSW) into renewable fuel can provide a means to avoid release of methane emissions from landfills into the atmosphere and as well a means to use MSW and hydrogen to provide clean liquid fuels/chemicals with lower carbon intensity than fossil fuel. We have analyzed the potential lifecycle greenhouse gas (GHG) reductions from conversion of MSW to methanol (MeOH) using gasification with hydrogen enhancement where the hydrogen is produced by electrolysis or by natural gas pyrolysis. Hydrogen enhanced MSW conversion increases the amount of methanol produced from a given amount of MSW by around a factor of two. We have given particular emphasis to the use of natural gas pyrolysis because of its potential for providing lower costs and its substantially reduced dependence on renewably-sourced electricity. MSW is pretreated to Refused Derived Fuel (RDF) before being fed to an illustrative case of entrained fluidized gasifier. A lifecycle analysis (LCA) is performed for this illustrative MSW gasification and methanol production process and the associated landfill methane avoidance. For the case of pyrolytic H_2, these results are for an illustrative fugitive methane emissions level of 1 %. The net GHG emissions impact of varying levels of fugitive methane emissions on net greenhouse gas emissions is analyzed and illustrates the importance of minimizing these emissions.

74d Conversion of Biomass-Derived Compounds into Jet Fuel-Range Hydrocarbons Using Hierarchical Zeolite Catalysts

Odiri Siakpebru, Lakshmiprasad Gurrala, Ana Colaco Morais

Short-term Sustainable Aviation Fuel (SAF) production strategies are being developed to convert fats, oils, lipids and starch sugars to jet fuel-range alkanes. However, starch sugars and vegetable oils are not abundant enough to meet the projected fuel demand. Lignocellulose is a renewable non-edible feedstock, which is highly available and can provide the required aromatics and cycloalkanes to SAFs. In this work, we report the direct hydrodeoxygenation (HDO) of lignocellulosic biomass into liquid alkanes and aromatics in the presence of 10% Ni/H-β zeolite as a catalyst at 270 °C, 3 MPa of H_2 pressure and 6h of residence time. Under these conditions, high yield in liquid alkanes (carbon yield of 25.0 ± 0.9 wt%), with high selectivity toward C_7-C_{15} cycloalkanes (15.9 ± 0.4 wt%) was observed. We further investigated the performance of the HDO process using 10% Ni loaded in hierarchical H-β (H-β-h) and HY (HY-h) catalysts and compared with their parent counterparts. An in-dept characterization of parental and hierarchical zeolites will be provided. The stability and recyclability of the catalysts will also be discussed.

74f Evaluating Pyrolysis Gas from Waste Plastics As a Fuel for Portable Power Generation

Topher Taylor, Brian Eggart, Daniel Kulas, Sarvada Chipkar, Jeff Naber, David Shonnard

Chemical recycling technologies, including pyrolysis, have gained recent popularity in the past decade to help close material flows in plastics recycling. These technologies seek to breakdown waste plastics into monomers, liquid fuels, or other useful chemicals. During fast pyrolysis, the plastic thermally degrades in an inert atmosphere at 500-600 °C and short vapor residence time (2-5 sec), creating three main products: wax, liquid, and gas. While the wax and liquid products have commercial markets, the pyrolysis gas is typically consumed internally for process heat. This product, which is composed of a mixture of C1-C6 alkanes and alkenes, has a lower heating value (LHV) equivalent to that of propane. To the best of our knowledge, no one has previously evaluated pyrolysis gas as a fuel source, nor measured the emissions of its combustion in portable power generation.

In this work, pyrolysis gas was generated from a novel liquid-pyrolysis process using military waste PE and PP feedstock. The composition of the pyrolysis gas was determined to be 25 wt% methane, 6 wt% ethylene, 29 wt% propylene, 24 wt% butene, 11 wt% pentene, and 4 wt% hexene using GC-MS and GC-FID. The collected pyrolysis gas was compressed into a receiving tank up to 300 PSIG and then used to fuel a portable power generator. Due to the complex nature of the fuel, a vapor-liquid equilibrium model was created using the Peng-Robinson equation of state and Rachford-Rice equation to model the changing composition in the receiving tank as a function of filling/draining cycles. We found that the heavier compounds in the fuel slowly accumulate in the liquid phase of the tank over several filling/draining cycles. Despite this change in composition, the LHV of the fuel was found to remain constant across 100 filling/drying cycles. In addition, emissions data (including CO, CO_2, O_2, Total Hydrocarbons, NOx) is reported from the combustion of the pyrolysis gas to evaluate the fuel quality. Finally, overall generator efficiencies were found to range from 20-25%, equivalent to commercial propane fuel.

74g Co-Gasification of Plastic, Coal Refuse, and Biomass for Hydrogen-Rich Syngas Production

Muhammad Aamir Bashir, Sittichai Natesakhawat, Jennifer Weidman, Nicholas Means, Fan Shi, Jonathan Lekse, McMahan Gray, Ping Wang

Plastic waste has increasingly become one of the most pressing environmental concerns. To mitigate plastic emissions effectively, extraordinary efforts are needed for its recycling and management. Gasification is a promising recycling approach to produce hydrogen and syngas suitable for sustainable energy and fuel applications. This study delves into the thermochemical conversion of mixed feedstock blends containing plastic, coal refuse, and biomass into hydrogen-rich syngas via steam gasification. Coal refuse or biomass (such as pine dust) is incorporated as a co-feedstock with low-density polyethylene (LDPE) to improve the handling of plastic waste and explore their synergistic effects on carbon conversion and product distribution across a broad spectrum of reaction temperature and feedstock composition. The investigation underscores the significant influence of process parameters, particularly temperature variations and blend ratios, on hydrogen yield and syngas quality. These factors are pivotal in determining the co-gasification efficiency and environmental sustainability of the process. LDPE demonstrates higher reactivity compared to coal refuse and biomass, due to its higher volatile content. Increasing LDPE content in the feedstock blends correlates with enhanced hydrogen yield, reduced carbon monoxide levels, and elevated concentrations of C_1-C_3 hydrocarbons, particularly methane. Within the investigated temperature range (800-1000 °C), positive synergistic effects on carbon conversion efficiency and hydrogen-rich syngas production are more pronounced at higher temperatures, resulting in an overall enhancement of gasification performance. This study contributes to advancing waste-to-energy techniques and highlights the importance of co-gasification of plastic and biomass in reducing carbon emissions and promoting a circular economy.

74h Gasification of Coal/Plastic Mixtures: Fundamental Studies in a Laminar Entrained-Flow Reactor

Afeez Kareem, Kevin Whitty

While literature reports various instances of plastic co-gasification with biomass/coal in fluidized bed gasifiers, research in coal and plastic gasification in entrained flow gasifier (EFG) is limited. Entrained flow gasification is of great interest considering its elevated temperature, flexibility of feedstocks fed, high efficiency, low cost of operation, low carbon emissions, and large-scale commercial applicability. EFG operates with liquid or slurries containing small solid particles at high pressure (3-8 MPa) and typically uses pure oxygen to partially oxidize the feedstock. Compared to other gasification types, it maximizes the carbon conversion rate, lowers methane content (enabling detachment of H_2 from hydrocarbon molecules), promotes atomization of slurries to increase the heat transfer rate, and typically runs in a slagging mode, i.e., at high temperature where the ash becomes molten.

To foster a circular economy, there is interest in the co-gasification of granulated plastics and pulverized coal in a pressurized, oxygen-blown, entrained-flow system. For this blended mixture, It is unclear how the introduction of plastics will affect coal conversion and the associated production of CO and CO_2. Relative to coal, plastics have a much higher volatile content (essentially 100%), and the volatiles are released at lower temperatures. With the addition of volatiles in the oxygen-rich portion of the burner, there is a question of whether oxygen that would otherwise be available for coal burnout will be consumed by the plastics volatiles (competitive oxidation).

To answer the research question, the University of Utah has lab-scale facilities to help understand the fundamentals of co-gasification, including a high-temperature laminar entrained-flow reactor (LEFR) and thermogravimetric analyzers, as well as analytical equipment including gas analyzers, an analyzer for total carbon and sulfur, and scanning electron microscopes. The experimental setup involved precise temperature control, plastic concentration, particle size, and residence time within the LEFR. Samples were prepared, loaded, and subjected to carbon conversion and syngas composition analysis. Findings revealed nuanced effects of temperature, plastic concentration, particle size, and residence time on coal-based carbon conversion and CO/CO_2 ratio. Elevated temperature, smaller particle size, and longer residence time enhanced carbon conversion, while plastic concentration notably increased the CO/CO_2 ratio. This research underscores the significance of temperature, plastic concentration, particle size, and residence time in shaping carbon conversion and syngas composition. The findings provide valuable insights into optimizing gasification parameters and pave the way for future investigations.

205a Production Marine Diesel Fuels from Olive Mill Wastewater Sludge

Foster Agblevor, Hamza Abdellaoui, Kamel Halouani

Valorization of agro-industrial residue such as olive mill wastewater sludge (OMWS) is very important to Tunisia and Mediterranean olive oil industry. Advanced methods of processing OMWS to convert it into marine diesel fuel and biochar will address both environmental pollution as well as reduce carbon emission. A thermochemical conversion method, such as fluidized bed pyrolysis was conducted at 450 C to produce biocrude oils from OMWS. The pyrolysis process produced large quantities of hydrogen which could be potentially used for hydrotreating the biocrude oil. The biocrude oil was suitable for direct use as marine diesel because of its low oxygen content, high HHV, and easy miscibility with very low sulfur diesel fuel. Hydrotreatment of the biocrude oil improved its fuel properties and could find a wide range of applications as diesel fuel. The biochar had very high carbon content and nutrients and could be potentially used for soil amendment in the olive growing areas. Thus, we have demonstrated that a potential waste material can be converted into very useful materials ensuring complete closure of the carbon circle.

205b Two-Stage Hydroconversion of Algal Oil to Sustainable Aviation Fuel (SAF)

Priscilla Manteaw, Adeniyi Lawal

Research into Sustainable Aviation Fuels (SAFs) continues to gain interest as an alternative fuel for decarbonizing the aviation sector. The US government's initiative, the SAF Grand Challenge, aims to spur research and industry towards the production of 35 billion gallons of SAF annually by 2050. Unlike other energy sources which require extensive infrastructure changes, SAFs adhere to ASTM specifications for jet fuels, making them compatible with current infrastructure. Amidst the biojet fuel technologies under development, leveraging microalgae shows promise due to their rapid proliferation rates and high oil yields. This study explores a novel pathway for converting algal lipids to SAF via a two-stage hydrodeoxygenation-isomerization process. In the first stage, hydrodeoxygenation is conducted to reduce the oxygen content, followed by the isomerization of linear alkanes to meet jet fuel specifications, in the second stage. Algal oil feedstock was extracted from concentrated algal slurry, and Fatty Acid Methyl Ester (FAME) analysis on algal oil shows that long fatty acid chains ranging from C_{13} to C_{20} are predominant. Therefore, for the hydrodeoxygenation step, analysis of the liquid product focused on the formation of these C_{13} to C_{20} hydrocarbons. Hydrodeoxygenation experiments were carried out on varying concentrations of algal oil in dodecane for different process variables. Catalysts employed were presulfided NiMo catalyst, a conventional hydrotreating catalyst, and precious metal-based catalysts: 1% Pt/Al_2O_3 and 0.5% Rh/Al_2O_3. Results indicated higher yields at lower oil concentrations, with the precious metal-based catalysts having a superior performance compared to the NiMo catalyst. Plans are underway to advance to the second stage using an isomerization catalyst, and evaluation studies will be conducted to identify the optimal process conditions for high product yields. Overall, the successful hydrodeoxygenation step validates the feasibility of microalgae-based biojet fuel through a two-stage approach and highlights the potential of commercial-scale production of SAF using microalgae.

205c Ozone-Cracking of Lipids to Synthesize Biokerosene and Its Combustion Performance in Diesel Engines

Junli Liu, Nathan S. Mosier

Net-zero emission is the globally admitted strategy to mitigate climate change. The United States will achieve carbon neutrality by 2050 by replacing the fossil economy with a bioeconomy through bio-based fuels, chemicals and materials. The transportation sector is a major GHG contributorin the United States. With passenger vehicles and small trucks with internal combustion engines being phased out by electric vehicles by 2035, heavy-duty engines, such as airplanes, still use liquid fuels for the foreseeable future. Various approaches have been investigated for biofuel synthesis from renewable sources to reduce GHG emissions. However, these technologies own several disadvantages, such as high energy consumption in production and refinery, low yields of target products, or harsh reaction conditions.

Herein, energy-efficient technology is desired to achieve net-zero or even negative carbon emissions. Lipids produced from crops, algae, or lignocellulosic sugar fermentation can be promising feedstocks for synthesizing low-carbon footprint fuel. Because of the high content of carbon-carbon double bonds, the current cracking refinery technologies can result in a high content of undesired aromatic molecules, resulting in undesired emissions, such as smoke, soot etc. Energy-efficient ozone cracking works as an alternative to the current thermal-based cracking process for low carbon footprint fuel production from lipids. Part of the ozonized products, carboxylic acids, reacted with bio-based alcohols to synthesize esters called biokerosene. Our results show that these products could be excellent alternatives to jet fuels or kerosene because of the excellent cold flow properties with cloud points as low as -70 °C. Alcohol types significantly influence viscosities and combustion heat. In addition, the products exhibit outstanding oxidation stability and thermal stability. The combustion performance of biokerosene and biokerosene-jet fuel blends in diesel engines was measured. The existence of oxygen facilitates combustion, resulting in low emissions of hydrocarbons and carbon monoxide and high carbon dioxide emissions. Moreover, biokerosene can reduce nitrogen oxides emissions. Therefore, our proposed approach provides an energy-efficient way to produce alternatives to petroleum-based jet, kerosene, and diesel fuels.

205d Biokerosene Produced from Ozone Cracking Lipids and Its Combustion Tests in a Diesel Engine

Junli Liu, Nathan S. Mosier

The utilization of fossil energy has resulted in global warming and other environmental issues. Diesel fuels account for about 17 % of total carbon dioxide emissions. Biodiesel, renewable diesel, and sustainable aviation fuel are solutions to reduce emissions for carbon sequestration. However, biodiesel use is constrained by poor low-temperature performance, making them unsuitable for cold winter seasons or jet fuel. On the other hand, renewable diesel is also challenged by the high operating temperatures and pressure, high energy consumption in production and fractionation, high demand for hydrogen sources, etc. An innovative production process was investigated to synthesize biokerosene from various alcohols and nonanoic acid, a primary product of the ozonolysis of oleic acid. The products exhibit excellent low-temperature performance because cloud points of the nonanoic esters can be as low as -70 °C. This super low-temperature performance showed the potential for replacing the winter-season diesel, kerosene, and jet fuels. Because of the saturation of the molecules, the products showed excellent oxidation stability and low greenhouse gas emissions. The energy densities of the products increased with the carbon atoms in the alcohols but were less than current biodiesel. The products generally had flash points higher than 90 °C, indicating safer handling and storage. In addion, the type of alcohol used in synthesis significantly influences on the physcial properties, such as viscosities, combustions of heat. The existence of oxygen in the fuel molecules facilitates complete combustion, resulting in low measured emissions of hydrocarbons and carbon monoxide and high carbon dioxide concentrations in engine exhaust. Furthermore, diesel engine exhaust from biokerosene combustion contained lower concentrations of nitrogen oxides than fossil diesel run in the same engine. Therefore, ozone-cracking of lipids provides an energy-efficient way to produce alternatives to petroleum-based jet, kerosene, and diesel fuels.

205e Catalytic Hydropyrolysis of Beetle-Killed Trees for the Production of Transportation Biofuels

Oluwanisola Makinde, Fernando Resende

In this study, we conducted catalytic hydropyrolysis experiments on beetle-killed trees such as Pine, Ash borer and Red bay. The experiments were performed using a micro pyrolyzer-Py/GC-MS setup. This study focused on the hydropyrolysis of beetle-killed trees in the presence of various heterogeneous catalysts like HZSM5, NiMo-HZSM5, and NiRe-HZSM5. We investigated the effects of different process parameters such as temperature, catalyst-to-biomass ratio, and hydrogen partial pressure on product yield. Our findings suggest that across the three feedstocks, temperature, pressure, and catalyst-to-biomass ratio play crucial roles in producing aromatics essential for transportation biofuel. The influence of catalyst acidity on product yields was also reported in this work and we observed that increasing the acidity of the catalyst positively influences hydrocarbon yield, irrespective of the feedstocks. Additionally, we examined the effect of acidity on the aromatic distribution of the product, revealing a more pronounced yield of C7 and C8 carbon numbers with increasing acidity. NiMo-HZSM5, and NiRe-HZSM5 have the potential to produce hydrocarbons for transportation biofuel as a result of the synergistic effect of metals doped on HZSM-5. Although we hypothesized that Re is a better alternative compared to Mo-containing zeolites because of its higher catalytic performance and stability compared to Molybdenum, the results we obtained with NiRe-HZSM-5 showed otherwise as we obtained only a small yield of alkanes compared to NiMo-HZSM-5. Additionally, we utilized response surface methodology, a blend of mathematical and statistical technique to optimize the various operating conditions for achieving the best yield. Our analysis revealed that the maximum bio-oil yield of 35% occurred at 570°C, 13.79 bar, and catalyst to biomass ratio of 13.9:1; an optimal aromatic yield of 28.3% was observed at 562°C, 13.79 bar, and catalyst to biomass ratio of 13:1 and finally, optimal alkane yield of 5.7% was achieved at 400°C, 13.79 bar, and catalyst to biomass ratio of 8.6:1

205f Oligomerization of Ethylene on Novel Ni-Siral Catalysts

Kristinoba Olotu, Fernando Resende

The adverse environmental impacts of fossil fuels have expedited the development of alternative energy technologies for the production of biofuels. Jet fuel molecules, such as linear and cyclic hydrocarbons can be produced from sustainable sources with the assistance of solid heterogenous catalysts. Such catalysts promote reactions that led to high yield and selectivity of linear and cyclic hydrocarbon compounds which are constituents of jet fuel. The oligomerization of ethylene is one of the main routes for the conversion of ethylene into hydrocarbon mixtures. The present study investigates the oligomerization of ethylene for the production of liquid hydrocarbons on novel Ni-SIRAL catalysts, with 4.9 wt. % nickel loading on several SIRAL supports (SIRAL 10, SIRAL 20, SIRAL 30, SIRAL 40, SIRAL 40 HPV and SIRAL 70 HPV which have pore volumes of 0.75, 0.8, 0.9, 1.0, 1.5 and 0.9 ml/g respectively). SIRALs [x%Al_2O_3: y%SiO_2] have a comparatively high level of surface acidity and large pore volume to their microporous and mesoporous support counterparts, which minimize coking and consequently catalyst deactivation. This study compared yields and selectivity for six catalysts (Ni-SIRAL 10, Ni-SIRAL-20, Ni-SIRAL 30, Ni-SIRAL 40, Ni-SIRAL 40 HPV and Ni-SIRAL 70 HPV) for the ethylene oligomerization process, with the aim to determine the most promising SIRAL support for increased liquid yield and product selectivity to linear and cyclic hydrocarbons. The experiments were carried out in a fixed bed tubular reactor at 200^0C and 40 bar. The maximum coke yield was 2.5 wt. % obtained with lowest pore volume catalyst (0.75 g/l), and the highest hydrocarbon liquid yield (36 wt. %) was achieved with the largest pore volume (1.5 ml/g). The liquid products obtained contain mainly C_4-C_{10} olefins.

205g Effects of Solvents and Upgrade on the Hydrothermal Liquefaction Products of Saccharina Latissima

Mayokun Oke, Fernando Resende

We converted the brown macroalga, Saccharina latissima using hydrothermal liquefaction in three solvent schemes (water-only, water-methanol co-solvent, and methanol-only), and the heterogenous catalyst, HZSM-5. We investigate the effects of these on the yields of product, as well as biocrude composition. We conducted the experiments in a batch reactor at 330°C (sub-critical water, supercritical methanol) and 1hour residence time. We found the highest biocrude yield (27.8 wt.%) in the catalytic run with methanol-only solvent, but the absence of water in this run led to the removal of the aqueous product phase (WSP). The impact of the zeolite catalyst, HZSM-5 in hydrothermal liquefaction appears to be solvent-dependent. The most significant impact of the zeolite catalyst was seen in the runs with water only, while methanol-containing runs only show marginal improvements in biocrude yield. We identified fatty acids and esters as the prevalent compounds in the analysis of the biocrude (43.9% - 69.1%), and their selectivities increased with methanol volume used as solvent (from co-solvent to only solvent). We also observed the occurrence of aliphatic chain alcohols in runs involving methanol, and these were not significantly present in water-only runs, with alcohol selectivity ranging from 1.0% (20% methanol) to 6.1% (100% methanol). The highest yield of hydrocarbons (0.8 wt.%) was obtained from the catalytic run with HZSM-5 and 20% methanol as co-solvent. We further subjected the obtained biocrude to upgrading in a pyrolyzer, and HZSM-5 catalyst proved most effective at the biocrude-to-catalyst ratio 1:5, producing 4.1 wt.% hydrocarbons yield, primarily composed of aromatics. We compared biocrude pyrolysis to that of raw macroalgae in the presence of HZSM-5 catalyst and found the quantitative yield and combined selectivities for alkenes, aromatics, and cycloalkanes from the direct pyrolysis of the macroalgae to be slightly lower than that of the biocrude, suggesting that pyrolysis of biocrude obtained from hydrothermal liquefaction of macroalgae could be a more efficient process for producing desired hydrocarbons than direct catalytic pyrolysis. These findings are promising to support the utilization of S. latissima as a potential substitute to produce transportation drop-in fuels.

73a Thickening of Supercritical Carbon Dioxide By Poly-1-Alkene Oligomers

Tomás Chávez-Miyauchi, Taniya Kar, Abbas Firoozabadi

Supercritical CO_2 ($scCO_2$) thickening is the most promising approach may be very effective in mobility control to increase sweep efficiency of storage in saline aquifers and in enhanced oil recovery (EOR) processes. Poly-1-decene (P1D) is an oligomer which is very effective as a $scCO_2$ thickener. The molecule is environmentally friendly and has very low adsorption to mineral surfaces. P1D can be dissolved in CO_2 from low to moderate concentrations. The oligomer of 20 repeat units can increase the relative viscosity of $scCO_2$ up to 6 times at 1.5 wt% concentration (Kar & Firoozabadi 2022). In this work, a series of poly-1-alkene molecules specifically engineered, based in monomers of 1-octene, 1-decene and 1-dodecene with different molecular weights, are examined for thickening of $scCO_2$. The goal is engineering of molecules that are even more effective than the poly1-decene oligomer with 20 repeat units (P1D-20) in our recent work. We observe a clear trend in maximum solubility at 35°C and 3500 psi where longer monomer chain lead to less solubility at the same molecular weight. Viscosity of the thickened $scCO_2$ is higher for molecules with longer monomer alkyl chain at the same molecular weight. Effect of pressure and temperature are also investigated. This is the first set of experiments evaluating the effect of poly1-alkene monomer chain length, concentration, pressure, and temperature in $scCO_2$ thickening.

Kar, T., Firoozabadi, A. (2022). Effective viscosification of supercritical carbon dioxide by oligomers of 1-decene. *IScience*, *25*(5), 104266.

73b Impact of Pressure on Thermodynamic Behavior of Waxy Model Oil with Methane

Bala Saishree Krishna Ala, Nagu Daraboina

This research delves into the impact of pressure on the thermodynamic behavior of waxy model oil. High pressure Differential Scanning Calorimeter (DSC) is utilized to understand pressure's effect on wax deposition and to determine wax appearance temperature (WAT) and precipitation under varying pressures. At higher pressures, the heavier molecules are more likely to remain dissolved, reducing waxy oil properties. Conversely, reduced pressure, typical in production and transportation, can induce precipitation of these components. The paper underscores the pressure dependency of oil's phase behavior, noting that high pressures can alter fluid phases, thereby influencing WAT. This behavior is crucial for understanding component interactions under different pressure conditions. The use of laboratory experiments, particularly Differential Scanning Calorimetry (DSC), is emphasized as a means to understand pressure's effect on wax deposition and to determine WAT under varying pressures. Understanding these dynamics is pivotal for the oil industry in developing effective strategies to address wax-related issues.

73k A Comparative Economic Evaluation of Treatment Versus Re-Injection of Produced Water in Texas Oil Extraction Operations

Matthew Kolade Ajakaiye, M. R. Riazi, Francisco Haces Fernandez

The escalating demand for energy resources underscores the significance of managing produced water, a byproduct of oil extraction, which presents critical environmental and economic challenges. Particularly in Texas, a leading oil-producing region globally, the treatment and disposal of produced water are fraught with complexities. This study conducts a comprehensive comparative economic assessment between treating produced water and re-injecting it into reservoirs post-oil extraction in Texas. Furthermore, it examines potential revenue streams and cost savings associated with each approach, including the sale of treated water for beneficial reuse and the avoidance of disposal fees through re-injection. Through an extensive review of literature, industry reports, and empirical data, this study aims to offer insights into the economic viability and sustainability of both treatment and re-injection strategies for managing produced water in Texas oil fields.

Employing a quantitative analysis approach, this research evaluates the costs linked with treatment and re-injection strategies, considering factors such as initial capital investment, operational expenses, regulatory compliance, and environmental impacts. Literature indicates that the volume of produced water could surpass that of the produced hydrocarbon by over tenfold during a reservoir's lifespan. Despite the advent of new technologies, traditional methods like gravity separation, hydrocyclone, and media filters, with a cost of $0.509 per cubic meter, are favoured for treating produced water for injection purposes. However, in regions like Iran with poor water quality, reuse options are limited, and efficient technologies are scarce. Mechanical Vapor Compression (MVC) and Multi-Effective Desalination (MED) technologies emerge as viable desalination options. According to findings, an MVC treatment system with appropriate pretreatment can recover 50% of produced water, representing the optimal water recycling option with a total treatment cost of $3.808 per cubic meter.

The research underscores the pivotal role of economic factors in determining the life cycle of oil fields, as the substantial volume of produced water and evolving environmental regulations significantly impact production, treatment, and disposal costs. The findings contribute to a nuanced understanding of economic trade-offs in produced water management within the Texas oil industry, thereby informing decision-making processes for policymakers, industry stakeholders, and environmental practitioners. Ultimately, the research aims to foster the adoption of more efficient and environmentally sustainable strategies for produced water management, aligning with broader goals of sustainable development and resource conservation in the energy sector.

73d Initiation Steam Cracking of Hydrocarbons By Hyperbranched Poly(amidoamine) (PAMAM) Initiator

Yunbo Yu, Tinghao Jia, Yu Ren, Jingdai Wang, Yongrong Yang, Congjing Ren, Yao Yang

The addition of initiators, possessing bonds with low bond dissociation energy, has been widely recognized as an effective method to promote the hydrocarbon cracking reactions. We synthesize the amidoamine-structured hyperbranched polymer (PAMAM) with the molecular weight of 12,500 as a novel macroinitiator, using ethylenediamine and methyl acrylate as raw materials. The experimental results show that PAMAM can significantly promote the cracking conversion and reduce the initial cracking temperature of hydrocarbons because the decomposition of PAMAM at relatively low temperature produces high concentration of active $\bullet NH_2$ and $\bullet CH_3$ radicals, which significantly increase the rate of hydrogen abstraction. To further investigate its initiation effect on hydrocarbons with different molecule structures, the steam cracking of three C6 hydrocarbons, namely *n*-hexane, *iso*-hexane and methyl-cyclopentane (MCP) were investigated experimentally in a stainless-steel tubular reactor, and the detailed kinetic reaction models and reaction process simulations are conducted by Reaction Mechanism Generator and Chemkin-pro software packages. The results confirm that the promotion efficiency of the initiator can be highly dependent on the cracking behaviour of hydrocarbons. The initiator will show higher promotion efficiency for hydrocarbon that is more prone to undergo sustained hydrogen abstraction after initiation. Therefore, the promotion efficiency of PAMAM follows the order of *n*-hexane > *iso*-hexane > MCP. This study will provide a direction for efficient production of light olefins from steam cracking of naphtha.

73f Catalytic Upgrading of Used Motor Oil for Marine Diesel Fuel Applications

Foster Agblevor, Hamza Abdellaoui

Used motor oil (UMO) is a waste product generated in large quantities in every country throughout the world. During normal use, impurities such as dirt, metal scrapings, water, or chemicals can get mixed in with the oil, so that in time the oil no longer performs well. Chemical impurities can include solvents, halogens, or saltwater. UMO is an environmental pollutant and is disposed of in various forms such as recycling, re-refining, and combustion to name a few. Because of its high sulfur content (1-3 wt%), it cannot be used as road transportation diesel fuel. We have developed a one-step catalytic conversion process that removes water, sulfur, and other impurities from the oil. The process reduces the viscosity and sulfur of the oils to meet the specifications of the international Marine Organization standards for marine diesel fuel.

73g Advanced Catalytic Process for Monetizing Heavy Aromatics Stream in Aromatics Complex to Increase Paraxylene Production

Qi Xu, Mohammad Alabdullah, Robert Hodgkins

A sustained growth in demand for petrochemicals has provided impetus to the on-going energy transition to alleviate the forecast erosion of gasoline value in the refining industry. Benzene and paraxylenes are among the valuable petrochemical products produced in a refinery aromatics recovery complex which typically also produces a reject heavy aromatics stream containing alkyl-bridged non-condensed multi-aromatics. The use of this heavy aromatics reject stream is limited to gasoline blending due to its composition, density, dark color, and high boiling point. In this context, Saudi Aramco has developed an advanced hydro-dearylation process, so called BTX-Max™ process, to upgrade this low-value reject stream over a zeolite-containing hydrocracking catalyst to significantly enhance high value mono-aromatics content.

The technological advantages of the BTX-Max™ process are multiple-fold: (1) dealkylate the di-aromatics into mono-aromatics; (2) lower the density of the stream; (3) improve the dark color to a much lighter one; (4) lower the average boiling point of the stream; and (5) process at relatively mild reactor conditions. Experiments have been carried out in a pilot plant testing unit with 72 grams catalyst under temperatures ranging from 200 °C to 400 °C and 15 bar pressure. Two-dimensional gas chromatography (2D-GC) data of the heavy aromatics reject stream and the product stream have been obtained to facilitate the hydrocarbon analysis. Simulation of the reactor effluent has been carried out to quantify the compositions and physical properties for the light aromatics and the unconverted heavy aromatics stream. Results have shown that the optimum temperature for the conversion of the multi-aromatics into mono-aromatics achieves is around 350 °C and the process yields a 21% increase in mono-aromatics. Results further confirm that mono-aromatic formation comes at the expense of biphenyl/di-aromatics conversion, with a 75% decrease observed. In addition to the technological advantages, the BTX-Max™ requires very few conventional equipment without hydrogen recycling, leading to very low CAPEX and OPEX.

In conclusion, a novel catalytic process, BTX-Max™, has been developed to monetize the low-value heavy aromatics reject stream in a refinery aromatics recovery complex. This process, under relatively mild conditions in the presence of a zeolitic catalyst and hydrogen, targets the alkyl bridge of the non-condensed alkyl-bridged multi-aromatics to form high-value mono-aromatics. The operation can recycle the mono-aromatics product stream back to the aromatics recovery complex to increase the production of benzene and para-xylene, and in so doing, capture the value created by this novel process.

73h Nonthermal Plasma Assisted Catalytic Crude Oil Upgrading Under Methane

Hua Song

Upgrading crude oil to acceptable lighter products to improve the downstream products' quality and to increase the topping distillate flows prior to reaching today's refineries secures economic benefits. The conventional upgrading processes include carbon rejection and hydrogen addition technologies. In these operations, thermal or thermal catalytic cracking converts large hydrocarbon molecules into smaller, more useful molecules by subtracting carbon while rearranging the chemical bonds of the original hydrocarbon molecules. The use of ultrapure hydrogen for crude oil upgrading is undesirable given their higher cost than other "H donor" sources such as methane. The use of methane to crude oil upgrading will unfold a chance to use wasted methane, which flares or emits into the air leading to pollution and uneconomical. Challenges associated with high temperatures and pressure for these processes have urged R&D to look forward to more sustainable and effective technologies. None of the one-size-fits-all technology exists for upgrading crude oil owing to the variability in its physicochemical properties from every oil field. This creates opportunities for novel technologies to process a large range of crude oils to achieve better process efficiency. It has been witnessed over the last few decades intensive efforts of both researchers and industries to improve crude oil upgrading processing in terms of process efficiency and economics, including optimizing operating conditions and developing more effective catalysts. Yet, upgrading of crude oils, even with commercial technologies available, remains capital-intensive processing units. With economics driving the crude oil development, novel technologies are being vigorously demanded to test and pilot for commercial-scale applications.

In-liquid non-thermal plasma is a novel but feasibly promising technology for crude oil upgrading under methane. Non-thermal plasma (NTP) is among the technologies rivaling the conventional thermal upgrading of crude oil. NTP is generated by ionized gas via electric discharge and is rich in vibrationally and electronically excited molecules along with other highly active species including radicals, ions, and free electrons. Given plasma operated far from thermal equilibrium, the activated species can be generated without a noticeable increment in bulk gas temperature. Dielectric barrier discharge (DBD) is most used for plasma-catalysis as this reactor design enables effective packing of the plasma region with optimized plasma-catalyst interactions. Such DBD systems are instantly safe on/off switch, flexible, and thereby offer the potential to be coupled with renewable, intermittent energy sources such as wind or solar energy for decentralized upgrading plant, which would enhance sustainability and diminish processing costs.

This research seeks to align the aforementioned challenges and opportunities for developing a series of cost-effective and coke-resistant catalysts and determining the optimal reactor configuration and reaction conditions for non-thermal plasma assisted catalytic upgrading of a variety of crudes under methane rich gas such as natural gas for producing high value-added liquid fuel/chemicals. In order to realize catalyst design in a logical method, isotopic labeled model compounds have been engaged to study the plasma-assisted catalytic reaction mechanism, the possible plasma-catalyst synergistic effect as well as the one occurring between methane and co-reactants, and the catalyst deactivation mechanism. The optimal reactor configuration was fabricated to conduct reactions at low temperature (25-100 ^{o}C) and atmospheric pressure. The designed catalysts have shown the strong ability to enhance the conversion of the reactants and the selectivity of desired product. A series of material characterization and control experiments have been conducted to study the reaction mechanism and catalyst deactivation. Quantum mechanical modelling and molecular dynamic simulation have also been employed as the complimentary tools to gain a better understanding of the involved plasma-catalytic mechanism at the atomic level, thus benefiting the catalyst rational design for performance optimization. It is also worth noting that upon the establishment of a comprehensive and diverse catalyst formulation database for the NTP catalytic process including all datasets in the public domain and the evolving datasets collected from this project, predictive models have been developed to recommend optimal catalyst formulations for specific chemical reactions, powered by artificial algorithms and machine learning (ML) techniques. Furthermore, ML algorithms have been employed to guide intelligent reactor design optimization, facilitate fundamental understanding of the inherently complicated NTP-catalytic reaction mechanism, and develop adaptive control strategies to manipulate and optimize emergent, collective behavior in NTP catalytic system. The combination of ML with NTP technology opens up new avenues for advancing our understanding and harnessing the potential of NTP in chemical and energy transformations.

This research is delivering a first of this kind technology which can efficiently convert low-cost hydrocarbons to high value-added liquid fuel/chemicals at near ambient conditions, thus enabling significant breakthroughs in the field of low-cost hydrocarbons upgrading. Furthermore, the outcomes derived from this project will make positive contributions to the methane emission reduction through getting it effectively converted to high value-added liquid fuel and chemicals along with other hydrocarbons abundant in Canada with minimized energy input in a format of electricity which can be generated from renewable sources, leading to net GHG emission reduction of at least 60%. The potential valorization capability of this technology is unquestionable considering the latest market situation. We can reasonably imagine that upon successful deployment of this technology at commercial scale, the economic and environmental benefits generated from this technology will help boosting the healthy growth of green economy in terms of new hires, increased advanced manufacturing capacities, and new devices as well as new catalyst materials and high valuable liquid fuel/chemicals using the most abundant natural resources in a clean way.

73j 3-Reactor Chemical Looping System for Point Source CO2 Capture and Subsequent Utilization for Liquid Fuel Production

Krutarth Pandit, Xiaofeng Jiang, Falguni Akulwar, Danwyn Aranha, Rushikesh Joshi, Liang-Shih Fan

Owing to the rising emissions and climate change, CO_2 capture and utilization from point sources is the need of the hour. Technologies that can, therefore, capture the CO_2 and then utilize it to form chemical products can lead to higher applicability. A novel 3-reactor chemical looping system is proposed with calcium ferrite ($Ca_2Fe_2O_5$) oxygen carrier material for the capture and utilization of point source CO_2 and its subsequent utilization in liquid fuel production. In the first reactor, a flue gas stream is inlet counter-currently on the oxygen carriers, wherein the $Ca_2Fe_2O_5$ oxygen carriers get carbonated to form $CaCO_3+Fe_2O_3$. The carbonated oxygen carriers are then transported to a second reactor, wherein they react with biomass to form high-purity syngas, which can then be integrated into the Fischer-Tropsch process for liquid fuel generation. The reduced oxygen carriers are then transported to a third reactor wherein the reduced oxygen carriers are reacted with air to regenerate the oxygen carriers back to $Ca_2Fe_2O_5$. Process simulation studies have been conducted to design the process and optimize the parameters. It is observed that the carbonation occurs better at a lower temperature, and the gasification reaction occurs with higher efficiency at a higher temperature. Therefore, heat integration strategies have been discussed to optimize process efficiency and achieve autothermicity. Further thermogravimetric experiments have been conducted to test the feasibility of the process.

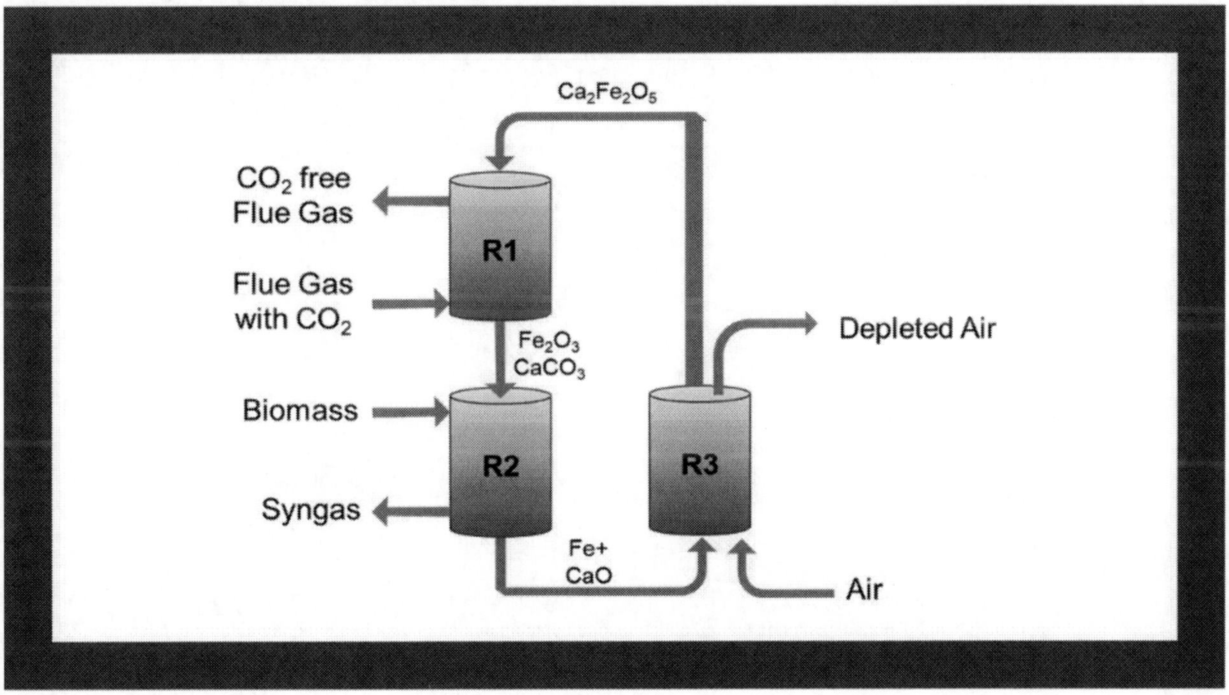

536a Enhanced Syngas Production with Reduced Tar Emissions: Optimization of Catalytic Biomass Gasification in Bubbling Fluidized Beds

Kshitij Tewari, Changle Jiang, Brandon Robinson, Debangsu Bhattacharyya, Jianli Hu

The goal of this DOE-funded research is to develop and test a cutting-edge, two-stage bubbling fluidized bed (BFB) gasifier using a hydrogen separation membrane that produces clean hydrogen from biomass. Reactor design innovation and optimization are critical to the advancement of biomass gasification technology, especially when it comes to producing cleaner syngas with an emphasis on hydrogen production. The configuration of the biomass gasification system is illustrated in Figure 1. We have optimized critical parameters such as feeding rate, minimum fluidization velocity, and flow-to-solids ratio to improve the system's performance. Furthermore, hot gasifier tests were carried out to examine how catalyst material and gas distributor design affected gasifier performance. The physical characteristics of the gas distributor and the effect of catalyst material on reaction kinetics and gas quality affecting flow distribution and residence time, were found to have a substantial impact on fluidized bed performance. The first catalyst bed is designed to produce syngas, and the second catalyst bed helps to decompose the tar. The ultimate objective of this research is to produce high-quality syngas with less than 1% tar concentration. Several reaction variables, such as the catalyst design, the gasifying agent, and the feeding rate oversight, make this outcome possible. By optimizing these variables, we want to reduce tar formation and produce syngas of the highest quality, thereby improving the dependability and efficiency of the gasification process. This work offers important new information about reactor design and operating parameters for biomass gasification research in the future.

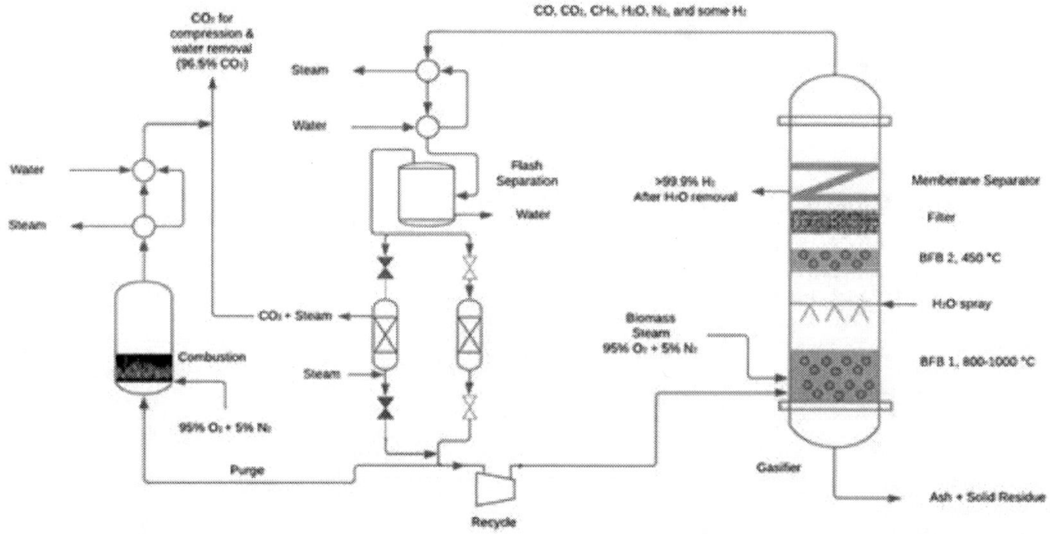

536b Optimizing the Combustion Synthesis of $FeAl_xO_y$ Catalysts for Microwave-Assisted Thermocatalytic Dehydrogenation of Fossil Fuels.

Zachary Chanoi, Laura A. Martinez Espinoza, Evgeny Shafirovich

The growing demand for hydrogen (H_2) requires the development of clean and energy-efficient technologies for its synthesis. Microwave-assisted thermocatalytic dehydrogenation of fossil fuels has demonstrated the potential to produce H_2 with high yield and selectivity, and simultaneously generate valuable nanostructured carbon byproducts. In prior work, iron-based alumina ($FeAl_xO_y$) catalysts for this process were made via solution combustion synthesis (SCS). However, the effect of SCS parameters on the dehydrogenation performance is not well understood. The present study investigates this by varying the SCS fuel, Fe:Al molar ratio, and heating mode. The results show subtle changes of these parameters can result in significant differences in the phase composition, specific surface area, and microwave absorbing properties of $FeAl_xO_y$, which all affect microwave-assisted dehydrogenation. Notably, H_2 selectivity can be increased from 30% to 74%. Statistical testing determined that the SCS fuel used was the most significant SCS parameter affecting dehydrogenation performance.

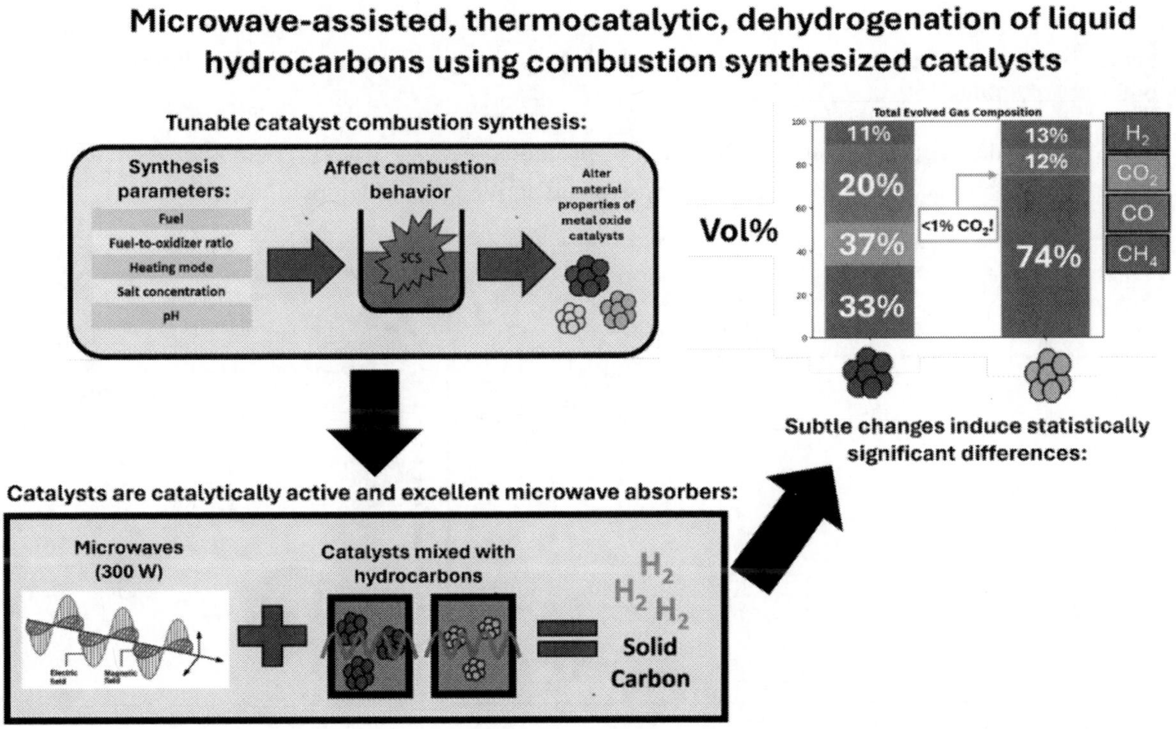

536c Microwave-Assisted Biochar Gasification: Effect of Metal Promoted Fe_3O_4, Water Cofeeding, and Biochar Composition on Syngas Production

Duy Hien Mai, Ashraf Abedin, Xinwei Bai, Dushyant Shekhawat, Pranjali Muley

The utilization of microwave-assisted gasification for biomass/plastic is a promising route toward clean energy production, contributing to a reduction in carbon footprint. This method facilitates the conversion of biomass into syngas with enhanced hydrogen (H_2) yield, surpassing conventional heating approaches. However, the gasification of biochar, a byproduct resulting from the initial rapid pyrolysis of biomass, appears as a rate-determining step in biomass gasification. To ensure high conversion efficiency, particularly in pilot or larger scales, the maintenance of high biochar reactivity is essential, which can be achieved by introducing catalysts to minimize biochar formation. Furthermore, given the susceptibility of biochar to microwave heating, gaining a comprehensive understanding of its behavior in the presence of microwave-active catalysts under microwave conditions is crucial to obtain valuable insights into the underlying mechanisms, thereby improving overall biomass gasification efficiency. In the previous study, magnetite (Fe_3O_4) was selected for microwave-assisted gasification for biomass/plastic due to its dual role as a catalyst and microwave absorber and demonstrated considerably enhanced hydrogen production. Herein, Fe_3O_4 is rationally modified with metal promoters and their synergistic effect toward biochar gasification performance is investigated. The data shows that metal promoted Fe_3O_4 shows a higher syngas yield than that of pristine Fe_3O_4 in biochar gasification. Additionally, the impact of water cofeeding and biochar derived from different biomass sources with varying compositions toward the performance outcomes are examined.

Figure 1. (a) Powder X-ray diffraction (PXRD) of Pt-promoted Fe_3O_4; (b) Correlation between ash content in various biomass samples versus the passive and active volatile temperatures.

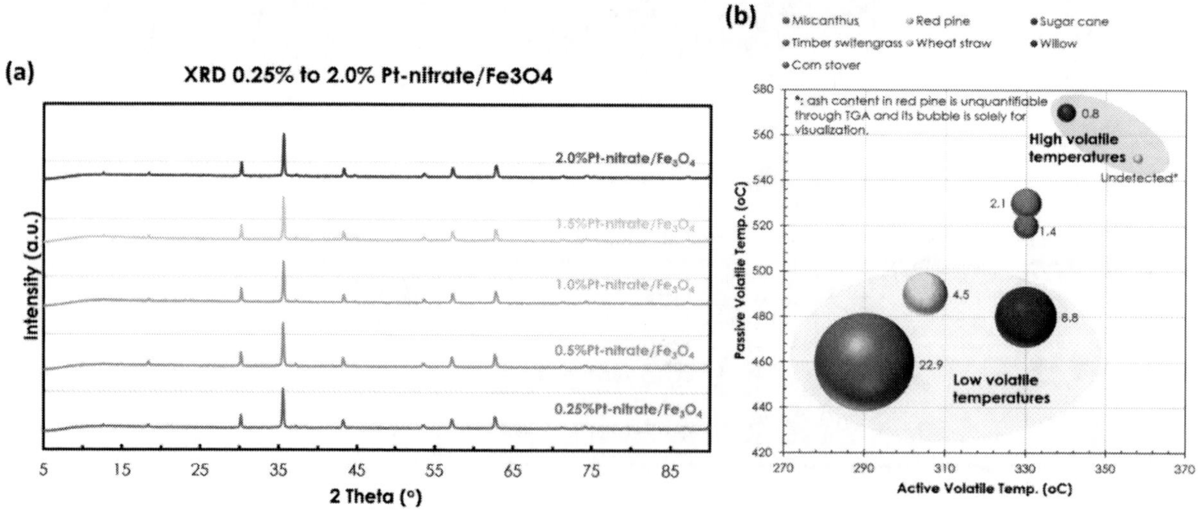

536d Fuel Flexible, Compact, High Pressure, Fuel Processor for H_2 Generation from Sustainable Feedstocks

Hani Hawa, Subir Roychoudhury, Seung-Hyuck (John) Hong

Precision Combustion, Inc. (PCI) has demonstrated a Catalytic Partial Oxidation (CPOX) reforming approach for low carbon intensity hydrogen (H_2) production from sustainable feedstocks, including Iso-Paraffinic Kerosene (IPK) and Hydro processed Esters and Fatty Acid based Synthetic Paraffinic Kerosene (HEFA-SPK) fuels. Preliminary test data indicated LHV-based reforming efficiency of >95% of the equilibrium value and stable operation without coking.

We have been developing CPOX and other application-specific fuel reforming approaches with both liquid fuels, such as jet fuel and diesel, and other sustainable fuels.

Our CPOX designs are based on a catalyzed short contact time, high surface area support (Microlith®), patented and manufactured by PCI. It reforms fuels to syngas in millisecond timeframes. No water or steam is needed for the fuel reformation yet coking is avoided. The short channel length, high cell density, metallic mesh support allows exceptionally high heat and mass transfer as well improved mixing with minimal pressure drop. These ensure short startup time and fast transient response. This also permitted a compact CPOX reforming. It operates at reaction temperatures and stoichiometries that are favorable for avoiding high temperatures while operating at high selectivity to syngas (i.e., carbon formation avoidance) and overcomes challenges with alternative substrates and catalysts. These characteristics make it suitable and attractive for integration with advanced power conversion devices, e.g., solid oxide fuel cells (SOFCs).

In this talk, we will report data from sub-scale high pressure testing with HEFA-SPK and compare results with equilibrium predictions as well as performance data from state-of-the-art syngas generation via other approaches.

536e Innovative Pressurized Diesel Reforming for Sustainable Hydrogen Productionusing Advanced PtRu-CGO Catalyst

Hyeone Lee, Yeongwoo Kim, Junwoo Park, Joongmyeon Bae

This study suggests a novel approach for producing hydrogen through the pressurized reforming of commercial diesel, leveraging the PtRu-CGO catalyst. Foundational experiments were performed with PtRu-CGO powder catalysts at conditions up to 9 bar and 800°C, achieving complete fuel conversion and a stable performance with an average hydrogen concentration of approximately 58.7 mol%. Additionally, the development of structured catalysts also showed that high efficiency and durability could be maintained under a gas hourly space velocity (GHSV) of 10,000 h^{-1}. Notably, at 6 bar, the system exhibited an extraordinary efficiency of around 75% over 300 hours of continuous operation. These results underscore the practicality of pressurized diesel reforming for hydrogen production with the PtRu-CGO catalyst, underlining its potential in the development and optimization of hydrogen energy systems and aiding the shift towards a hydrogen-based economy.

536f Modeling of Fluidized Bed Sorption-Enhanced Biomass Gasification for Hydrogen Production Using Aspen Plus

Jieun Kim, David Ray Wagner, Kevin Whitty

Sorption-enhanced gasification (SEG) of biomass represents an innovative approach for hydrogen production. This process utilizes renewable biomass and integrates CO_2 capture within the gasification step, enhancing hydrogen yield and purity. A comprehensive understanding and optimization of the process through simulation is essential to advance SEG towards commercial viability. Process modeling via Aspen Plus offers detailed insights into the operational parameters and system integration challenges. This model will facilitate developing an efficient, scalable, and economically feasible SEG process.

This study aims to create a comprehensive simulation that includes all stages of the SEG process, from biomass feedstock to steam gasification, limestone calcination, and transfer of the sorbents within a dual-fluidized bed configuration for CO_2 sorption. By detailed modeling of the SEG process, this work seeks to optimize the operational parameters and enhance the hydrogen production efficiency. Considering the Department of Energy's Hydrogen Shot initiative, which targets a significant cost reduction for clean hydrogen, this detailed modeling effort aims to refine operational parameters for improved hydrogen yield and economic viability, thereby advancing the scalability of SEG technology.

The Aspen Plus simulation is developed in several key steps. The feedstock characteristics, including composition and moisture content, are defined based on typical biomass materials. The fluidized bed reactor within Aspen Plus is customized to reflect the specific conditions and reactions, including sorbents for in situ CO_2 capture. This will involve incorporating kinetic models derived from lab-scale data to represent the gasification of biomass and sorption reactions of limestone. The simulation will also include sensitivity analysis to explore the impact of various operational parameters (e.g., temperature, pressure, feedstock, and solid circulation rate) on the efficiency of hydrogen production and the overall system performance.

The Aspen Plus model overviews the SEG process, identifying key factors influencing hydrogen yield, purity, and production cost. The simulation will identify optimal operational conditions that maximize hydrogen production efficiency while minimizing energy and material inputs. The sensitivity analysis is anticipated to highlight critical process parameters that significantly impact system performance, offering pathways for process optimization. This modeling effort will conclude recommendations for process design and operational strategies that enhance the scale-up application of hydrogen production through SEG.

Building on this modeling work, future research will explore advanced simulation techniques to incorporate dynamic process control and real-time optimization capabilities. The experimental data from the 200 kW pilot scale dual-fluidized bed system at the University of Utah will be used for model validation. Additionally, extending the model to include life cycle analysis and environmental impact assessment will provide a comprehensive evaluation of the sustainability of the SEG process. Ultimately, this research will contribute to developing a robust, efficient, and economically competitive SEG process for hydrogen production, supporting the transition to a more sustainable energy system.

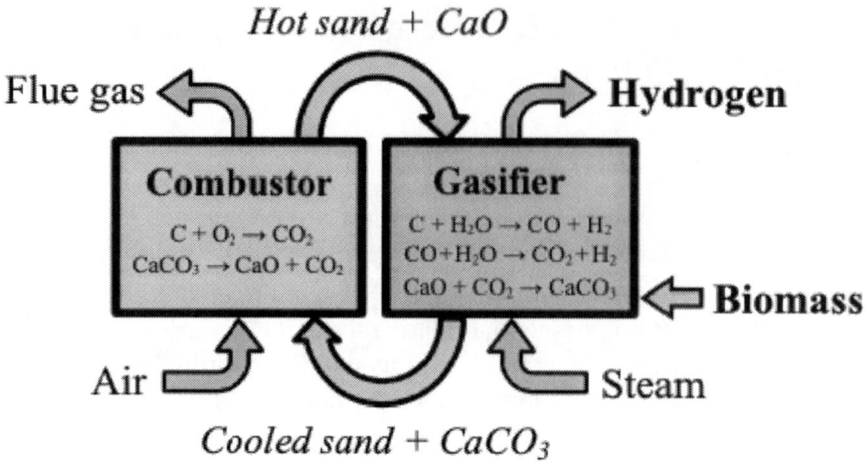
Hot sand + CaO
Flue gas
Hydrogen
Combustor
$C + O_2 \rightarrow CO_2$
$CaCO_3 \rightarrow CaO + CO_2$
Gasifier
$C + H_2O \rightarrow CO + H_2$
$CO + H_2O \rightarrow CO_2 + H_2$
$CaO + CO_2 \rightarrow CaCO_3$
Biomass
Air
Steam
Cooled sand + $CaCO_3$

536h Towards Circular Economy: Optimisation of Co-Culture Fermentation for Biohydrogen Production from Flour Mill Waste.

Akanksha Jain, Eshaan Agarwal, Gobinath Rajagopalan

The escalating global energy demand, coupled with the urgent need to mitigate the environmental impacts of fossil fuel reliance, drives the exploration of sustainable and renewable energy sources. Hydrogen [H2], a gaseous fuel, is a promising alternative to fossil fuels. It has approximately 2.75 times higher energy density than other gaseous fuels such as methane and CNG. Additionally, unlike other fuels, on combustion, it produces only water, making it a zero-carbon emitter. Among various methods for hydrogen production, biological methods are environmentally friendly as they are not dependent on fossil fuels and require much less energy and heat, unlike chemical methods. Most importantly, they can utilise biomass as a feedstock for hydrogen production. Microorganisms belonging to *Clostridium sp*. exhibit an interesting mechanism known as dark fermentation by which they can anaerobically utilize carbohydrates like glucose to produce hydrogen through the Acetic acid and Butyric acid pathway (ABE). However, to commercialize this hydrogen as fuel, it needs to be cost-competitive.A key factor limiting cost-effectiveness is the reliance on expensive, often food-grade feedstocks. The substrate is found to be contributing to 60% of the total bioprocess cost. Therefore, there is a need to explore cheaper non-food based substrates and test their suitability for fermentation.

Flour mill waste (FMW), a globally abundant and underutilized by-product of grain processing, presents an attractive low-cost and non-edible feedstock for biohydrogen production. On a global scale, a significant proportion of the human diet is filled with cereal grains such as wheat, rice, barley, maize, sorghum, millet, oat, and rye. According to the FAO (2016), global production of these crops is nearly 2577.85 million tons. For the production of refined flour, bran, and germ, a proportion of grain is removed and thus, nearly 8% of the production is wasted in the milling process. Its substantial carbohydrate content, primarily in the form of starch, provides a rich energy source for microbial conversion. However, the recalcitrant nature of starch poses challenges for direct fermentation by single microbial strains. This often results in inefficient substrate utilization, limited hydrogen yields, and the need for complex and costly pre-treatment processes, which not only take away from its sustainability but also pile on costs, making the biohydrogen less appealing and feasible. FMW is a waste that is generally an environmental nuisance mostly burnt leading to pollution. Thus its valorization for energy production also enables a safe and proper waste disposal methodology.

This study addresses the limitations of FMW through an innovative co-culture fermentation strategy designed to optimize both FMW hydrolysis and subsequent biohydrogen production. Initial characterization of FMW revealed a diverse sugar profile, including starch (50-70%), glucose (15-20%), maltose (15-20%), xylose (1-5%), and arabinose (1-5%). A comprehensive screening of saccharolytic Clostridia strains, including *Clostridium beijerinckii* G117, *Clostridium butyricum* NCIM 2825, *Clostridium acetobutylicum* NCIM 2337, and *Clostridium sp.* BOH3, identified *Clostridium beijerinckii* G117 as the most promising candidate. This strain demonstrated the ability to utilize a diverse range of sugars (including starch, glucose, maltose, xylose, and arabinose), achieving high cellular growth (1.07-1.78 OD600)and hydrogen yield (0.81-1.92 l/l) , making it well-suited for FMW-based fermentation.

Direct fermentation of FMW as a feedstock (10-100 g/l) in minimal media was carried out using *Clostridium beijerinkii* G117. A maximum hydrogen yield of 2.7 l/l was obtained from 50g/l FMW. It was found that increasing FMW beyond 50 g/l, hydrogen production and cellular growth declined.

Also, 50-70% of the FMW was unutilized in all the cases, leading to feedstock wastage. Despite the strain showing the capacity of amylase production (1.15U/ml), the hydrogen yield remained significantly low despite the available sugar in the fermentation broth. It was suspected that the strain is under severe metabolic burden due to performing both saccharification and hydrogen production, therefore hindering its performance. Therefore, pre-treatment for saccharification of starch residues to reducing sugar, for easy assimilation by C.G117 was considered.

A systematic evaluation of various pre-treatment strategies was conducted to improve the accessibility of complex carbohydrates in FMW. These included physical methods (autoclaving, milling), chemical methods (acid and water hydrolysis), and enzymatic hydrolysis using both commercial α-amylase and a *Bacillus sp.* strain with high amylase production potential. A particular focus was placed on identifying both effective and cost-optimized hydrolysis techniques. The introduction of the *Bacillus sp.* strain significantly improved the saccharification efficiency of FMW, offering a more economical and sustainable solution compared to solely relying on commercial enzymes. Detailed analysis of the pre-treatment methods revealed that a combination of moist heat treatment followed by enzymatic hydrolysis provided the optimal balance of sugar release, cost-effectiveness, and environmental sustainability. We found that *Bacillus sp.* could produce amylase units up to 7U/ml and effectively hydrolyze starch in the supernatant by nearly 97%. Saccharification of high FMW concentration (60-150g/l) by *Bacillus sp.* followed by fermentation by C.G117 yielded high hydrogen (3-3.6 l/l) with a shorter lag phase in comparison to direct fermentation improving the hydrogen production rate as well.

Following pre-treatment optimization, a consolidated bioprocess (CBP) approach was implemented to streamline the process. The co-culture system, comprised of the selected C.G117 and *Bacillus sp.*, was further refined using Response Surface Methodology (RSM) with Central Composite Design (CCD). Key parameters such as inoculum ratio, FMW concentration, and pH were systematically adjusted to determine the conditions that maximize biohydrogen yield and substrate conversion efficiency. The optimized CBP achieved remarkable success, yielding over 4 L/L of biohydrogen with high FMW conversion rates (70-90%). This outcome represents a substantial improvement over direct fermentation methods, clearly demonstrating the effectiveness of the co-culture strategy in overcoming the challenges associated with utilizing complex biomass. Detailed analysis revealed that the co-culture system facilitates a synergistic relationship. The Bacillus strain effectively hydrolyzes FMW starch, providing readily accessible sugars for enhanced hydrogen production by the Clostridium strain.

This research firmly establishes FMW as a valuable feedstock for biohydrogen production. The developed co-culture system offers a robust and efficient platform for waste valorization. It generates clean energy while simultaneously addressing sustainable waste management practices. Importantly, this work aligns with the principles of a circular economy, where waste streams are transformed into valuable resources, fostering environmental sustainability and resource efficiency.

596a Design Study for Hydrogen Production through Fluidized Bed Heated Steam Methane Reforming (SMR) Combined with Chemical-Looping Combustion (CLC)

Tobias Pröll, Anders Lyngfelt

Chemical-looping combustion (CLC) of gaseous fuels provides concentrated CO_2 while avoiding the energy penalty of gas-gas separation. The fuel is oxidized in direct contact with an oxygen carrier material in the fuel reactor while the circulating oxygen carrier is re-oxidized with air in the air reactor. Steam methane reforming (SMR) requires relevant heat transfer rates to heat the tubular reformer reactors. If such reformer tubes are designed to be placed in a hot fluidized bed for heat supply, the heat for reforming can be supplied from CLC of additional methane and the off-gas from the pressure swing adsorption (PSA) unit isolating the H_2 product, ideally allowing for 100% CO_2 capture while achieving similar or even higher energy efficiency than common SMR.

The process according to Figure 1 has been modelled in the steady state simulation environment IPSEpro with a customized model library for high temperature gas-solid processes. The necessary gas and solids flow rates are determined and transferred to a bulk design geometry of the fluidized bed system. The fuel conversion performance of the SMR reactor and of the CLC system are set based on the results of previous detailed studies on certain parts of the overall process.

It turns out that relatively high solid circulation rates are needed to keep the temperature differences between the air reactor and reformer heat exchanger reasonably small. A design solution is developed to achieve sufficient solids exchange and still keep the turbulences around the heated reformer tubes reasonable. The geometry is inspired by industrial fluidized bed heat exchanger practice. Pressure balance calculations give some initial confidence about the fluidized bed system design.

In conclusion, despite the high energy-efficiency potential of the CLC-SMR combination, practical implementation of the fluidized bed system comes with challenges. A rational approach towards a possible design has been presented and requires fluid dynamic verification in a cold flow model as possible a next step.

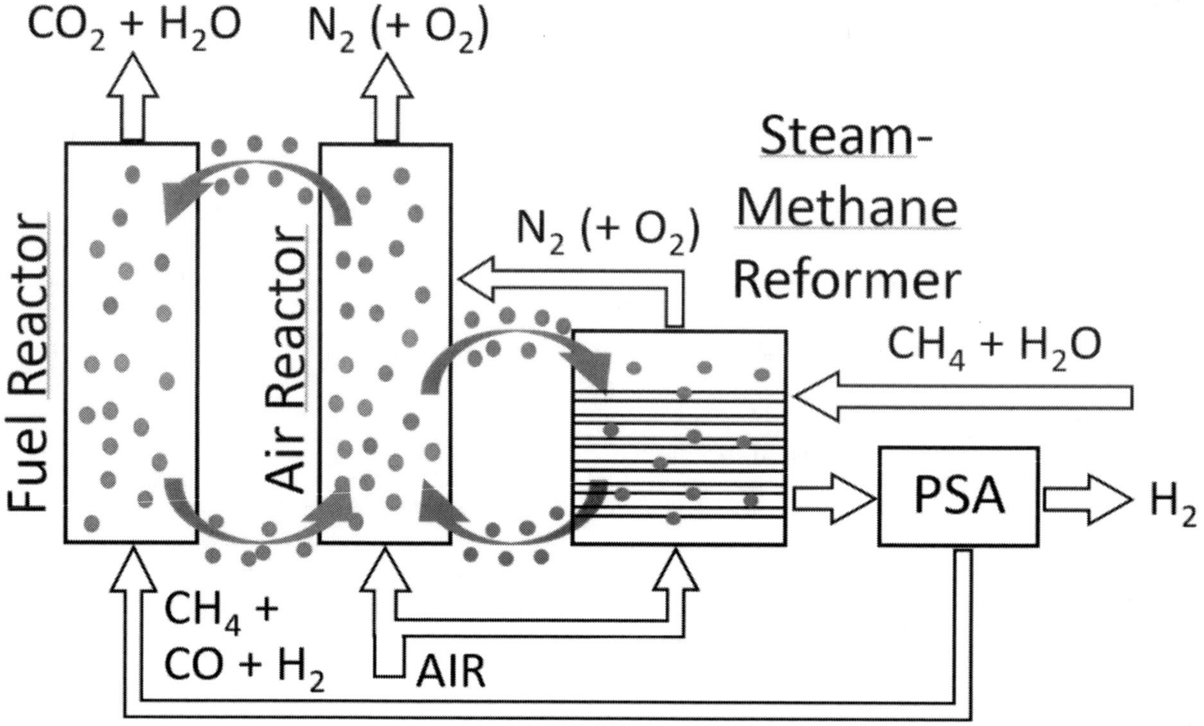

CO2 + H2O
N2 (+ O2)
Steam-
Methane
Reformer
N2 (+ O2)
Fuel Reactor
Air Reactor
CH4 + H2O
PSA
H2
CH4 +
CO + H2
AIR

596b Methane Dry Reforming over Ni Catalysts on High Oxygen Conductivity Supports

Xuan Pham, Janet Baffoe, Corey Leclerc

Dry reforming of methane process (DRM, CH_4 + CO_2 2CO + $2H_2$) converts methane and carbon dioxide, two major greenhouse gases, to syngas with an equimolar H_2/CO ratio. Syngas from DRM is ideal for Fischer-Tropsch synthesis to produce long-chain hydrocarbons and high-value chemicals. DRM is a potential process for the valorization of biogas or natural gas with high CO_2 content. Despite the positive environmental impacts, DRM faces a great challenge due to the rapid catalyst deactivation by the accumulation of carbonaceous species on active sites. Therefore, developing a highly stable catalyst system with strong coke resistance for long-term operation is necessary for the commercialization of DRM.

This research investigates the effect of supports possessing high oxygen conductivity on Ni-catalyzed dry reforming of methane. Ni-based catalysts are promising for DRM because of their relatively high catalytic activity and affordable costs, but they are easily deactivated by carbon deposition. Materials with high oxygen ion conductivity, such as Ceria (CeO_2), Gadolinium Doped Ceria (GDC), and Yttrium Stabilized Zirconia (YSZ) can provide oxygen ions for the metal to oxidize the deposited carbon and improve resistance to carbon deposition. These supports can form and eliminate oxygen vacancies, facilitating the adsorption and dissociation of CO_2, thus improving reforming activity. In this work, four support materials, which are very different oxygen ion conductivities, $Al_2O_3 < CeO_2 <$ YSZ < GDC, were evaluated for their activity, stability, and coke resistance. The kinetics of these nickel catalysts are also assessed by evaluating the methane activation energy and turnover frequency (TOF). The catalysts are characterized using various analytical techniques to understand the correlation between the catalyst's performance and their physicochemical properties. The long-term stability tests at 750°C in 12 hours show the deactivation in all catalysts but with different rates. The most stable catalyst is one with the highest oxygen mobility (Ni/GDC), while the catalyst lowest in oxygen conductivity (Ni/Al_2O_3) is the least stable. Carbon deposition, low in other cases, but is observed by scanning electron microscopy and temperature-programmed oxidation in the case of Ni/GDC. The structure or nature of carbon accumulated on the catalysts' surface after the DRM reaction is also clarified by Raman spectrometry. Other factors contributing to carbon formation and affecting the role of oxygen conductivity, such as active metal particle size, surface area, and acidity/basicity of support materials, have been also considered.

596i Non-Thermal Plasma Catalytic CO_2 Hydrogenation over a Hydroxyapatite Supported Ni Catalyst: Effect of Severe Process Conditions on CO Production

Esma Ines Achouri

*In the current work, the catalytic performance of a hydroxyapatite-supported nickel-zirconia catalyst, in a microwave plasma reactor for the CO_2 hydrogenation to carbon monoxide was studied. The 10 wt% Ni-10wt% ZrO_2/HAp was prepared by wetness impregnation method, dried at 180 °C for 18 hours and calcined at 500°C for 3 hours. The main factors influencing the catalytic performance are the H_2/CO_2 molar ratio and Power (W). The highest carbon yield ((CO_2 conversion * CO selectivity)/100) were 83% under H_2/CO_2 to 2:1, Power=2.25 kW, and GHSV=80,00 mL.gr^{-1}.hr^{-1} while maintaining 9% energy efficiency which to the best of our knowledge has never been reported in the literature. In this research, we embarked on a novel catalyst formulation for the CO_2 hydrogenation reaction, tested a microwave discharge for the reaction which has received less attention in the literature, and operated at very high GHSV levels which is suitable for industrial production.*

Highlights:

- The synergy between plasma and catalyst increases the CO selectivity.
- A novel bio-based catalyst under a microwave plasma discharge provides 83% CO_2conversion and 99.96% CO selectivity.
- The optimized H_2/CO_2ratio is 2.35.

· GHSV at very high ends does not affect the CO_2 conversion and CO selectivity.

596e Novel Metal Carbide (MXene) Based Catalyst for Industrial-Scale Electrochemical Production of Hydrogen

Yue Wu, Xiaopeng Liu

Hydrogen (H2) is considered as the most promising clean energy source to replace conventional fossil fuels. Despite the exceptional catalytic activity of Pt-group metals, the high cost and relatively less durability of these noble metal-based catalysts remain one of the key bottlenecks for sustainable hydrogen production at the industrial scale. In this report, we synthesized a new type of 2D metal carbide (MXene). By loading metal onto the surface of MXene, we developed a highly effective and stable catalyst for alkaline hydrogen evolution reaction (HER). The optimized catalyst exhibited a small overpotential of 62.58 mV at 10 mA/cm2 and a low Tafel slope of 44.34 mV/dec. At high current density of 100 mA/cm2 and 1000 mA/cm2, the low overpotentials of 191.18 mV and 407.72 mV were achieved, significantly outperforming the commercial Pt/C electrode. Under both current ranges, our catalyst exhibited excellent stability as 500 h at 10 mA/cm2 and for 100 h at 1000 mA/cm2 without any obvious degradation. In flow cell tests, by pairing up with Ni Foam, our catalyst also required much lower cell voltage than commercial Ni Foam||Pt/C and maintained ~ 100% H2 faradic efficiency over the 15h continuous tests from 100 to 400 mA/cm2. X-ray absorption near edge structure demonstrated a temperature induced redistribution of metal on MXenes from single sites and large nanoparticles to uniform small nanoparticles. Density functional theory calculation, combined with X-ray photoelectron spectroscopy and Extended X-ray Absorption Fine Structure Spectroscopy, revealed that the HER activity was sensitive to the coordination environment of metal on MXene and the metal-carbide interface might be the active site for HER chemistry. This work sheds light on structure design and utilization of metal-support interaction (MSI) in MXene-supported metal catalysts for the efficient electrochemical water splitting at industrial scale.

596f Membrane-Based Hydrogen Separation from Syngas Towards Refinery Decarbonization

Ragad Aldilaijan, Osamah Siddiqui, Stephen Paglieri, Minseok Bae, Alaa Albasry, Mohamed Draze, Abdulbari Alqarni, Ahmed Naimi, Aadesh Harale

Hydrogen (H_2) production is a steppingstone towards combating climate change by transitioning to more sustainable energy sources. Currently, H_2 is dominantly produced through steam methane reforming (SMR) process, which is considered to be an intense CO_2 emitting process. According to the International Energy Association (IEA), as of 2023, the average estimated CO_2 emissions intensity associated with H_2 production through SMR is around 10-14 kg CO_2-eq/kg H_2. Development in carbon capture and storage technologies can reduce the CO_2 emissions intensity down to 5-8 kg CO_2-eq/kg H_2 (with a CO_2 capture rate of around 60%), and to 0.8-6 kg CO_2-eq/kg H_2 (with a capture rate above 90%). Carbon Capture and Utilization (CCU) are used to produce low-carbon H_2. However, such process modifications are expensive to develop and operate, creating an obstacle in the deployment of low-carbon hydrogen from hydrocarbons.

Moreover, the final step of hydrogen purification through Pressure Swing Adsorption (PSA) or cryogenic distillation post the SMR unit contributes significantly to the overall cost of H_2 produced. Comparatively, the operation of palladium-based membranes requires lower cost and energy, in addition, its high permeability and selectivity for H_2 makes it exemplary for hydrogen purification.

This study investigates the use of palladium-based membranes for H_2 separation after different process steps with multiple heat recovery options to analyze the overall system performance under varying operation conditions. Aspen Plus V.12 is utilized to model the proposed configurations whereas the membrane-based H_2 separation model is developed in conjunction with lab-scale experimental results. The comparative performance of each modelled configuration and the progress of the palladium-based membrane technology at Saudi Aramco's Research and Development Center will be presented.

596g Noncatalytic Solid-State Hydrolysis of Sodium Borohydride with Sodium Metaborate Hydrate to Form an Isomorphous Hydrates

Hyun-Tae Hwang, Savannah Murt, Garrett Matheny

A method for safe and efficient hydrogen generation has been heavily researched in the past decades as a sustainable energy source for a variety of applications. Among hydrogen storage materials, sodium borohydride ($NaBH_4$, SBH) has been extensively studied due to its relative inexpensiveness and high hydrogen content (10.8 wt%). Traditional methods of generating hydrogen from SBH include thermolysis and hydrolysis. Unfortunately, thermolysis of SBH requires high temperatures (>300°C) even in the presence of a catalyst, while hydrolysis requires an efficient catalyst and excess water. There are also safety concerns with hydrolysis due to the possible unexpected generation of hydrogen when SBH and water are stored together. Ultimately, both hydrolysis and thermolysis present distinct challenges.

Recently, we have proposed a new process: thermal hydrolysis of a solid-phase mixture of SBH and sodium metaborate tetrahydrate ($NaBO_2 \cdot 4H_2O$, SMB4H). The thermal decomposition of SMBH releases water that reacts with SBH to produce hydrogen. With this, safety issues arising from self-hydrolysis pose no threat, as solid-state SBH-SMB4H mixtures are stable at ambient conditions. Since a high-density fuel is suitable for practical use of this approach, the characteristics of thermal hydrolysis of SBH were systematically investigated using a compressed mixture in the form of pellets in this study. In addition, this approach was investigated over a wide range of SBH/SMBH ratios, temperatures, pressures, pretreatment temperatures, and storage periods. During the process, rapid hydrogen evolution was observed around 150°C, and a maximum overall H_2 yield of 5.76 (2.4 H_2 equivalent) was obtained at 200°C for an SBH:SMB4H ratio of 1:0.6. The only products of the reaction are sodium metaborate and hydrogen, which makes this process more sustainable, indicating that this proposed method is promising for use in hydrogen storage for various applications including energy and commercial sectors.

596h Self-Sustained Hydrogen Production from Ammonia Decomposition

Amr Ibrahim, Sung Hwan Park, Saman Emami Gerami, Jennifer Naglic, Jochen Lauterbach, Miao Yu

Ammonia is a promising carbon-free energy carrier that will enable safe and viable use of hydrogen as a clean and efficient energy source. A self-sustained system for hydrogen production from the thermal decomposition of ammonia with a catalyst is highly required in the countryside, remote or emergency areas where external heat is hardly supplied. Herein, a self-sustained, compact reactor for ammonia decomposition was fabricated to provide a continuous production of hydrogen. Once initiated, the system is fed only with pure ammonia that is fully decomposed by a Ru-based catalyst to hydrogen and nitrogen gas mixture (75% H_2 and 25% N_2). Part of the decomposition product is mixed with air and fed back to the reactor where hydrogen is catalytically oxidized in an optimized palladium hydrogen burner, providing the required energy for ammonia decomposition. Different configurations of the palladium catalyst were investigated to provide a uniform heat distribution in ammonia catalyst bed and a fast drainage of the produced water vapor. Self-sustained ammonia decomposition was successfully demonstrated for more than 6 h, with approximately 72% of the produced hydrogen recoverable, compared to the maximum of 80% predicted by ASPEN HYSY. Steady state operation revealed negligible NO_X concentrations in the exhaust. Ammonia preheating and reactor insulation greatly affected ammonia decomposition efficiency. This research work will provide insights into the development of self-sustaining ammonia decomposition systems equipped with hydrogen selective membranes for pure hydrogen extraction.

596c Potential of Ni-Supported Upgraded Slag Oxide Pellets Towards Higher Than Atmospheric Pressure Syngas Production from Methane Reforming

Muhammad Irfan Malik, Nicolas Abatzoglou, Esma Ines Achouri, Juan Pablo Diaz Ramirez

This study explores the H_2-rich syngas production under varying pressures (1-10 atm) at 850 °C by dry autothermal and tri-reforming of methane over Ni pellets prepared from metallurgical residue. The influence of process parameters such as reaction pressure, temperature, feed ratio, gas hour space velocity, and role of O_2 circulation across the bed on the conversion of methane and syngas selectivity were investigated. The Ni-supported upgraded slag oxides (Ni-UGSO) pellets consistently showed more resistance to coke formation due to dispersion of Ni in Al, Fe and Mg-based spinels and silicates. The latter influenced positively the activity and catalyst robustness. CH_4 conversions and syngas yields close to equilibrium were reached at pressures between 1 to 10 bar. Typically, a 78% steady CH_4 conversion was obtained for 48 hours during dry autothermal reforming at 10 bar, 850 °C under molar ratios of CH_4/O_2 = 2 and CH_4/CO_2 = 3. Similarly, the addition of steam to simulate a tri-reforming condition (CH_4/O_2 = 2 and $CO_2:H_2O$ 1:1) at 10 bar and 800 °C has increased the H_2/CO ratio and minimized the coke deposition over longer time on stream. These low-cost and higher-performance Ni-UGSO pellets are a potential candidate for successful and sustainable industrial syngas production, i.e., biogas-to-liquid (BTL) fuel synthesis.

Some preliminary results

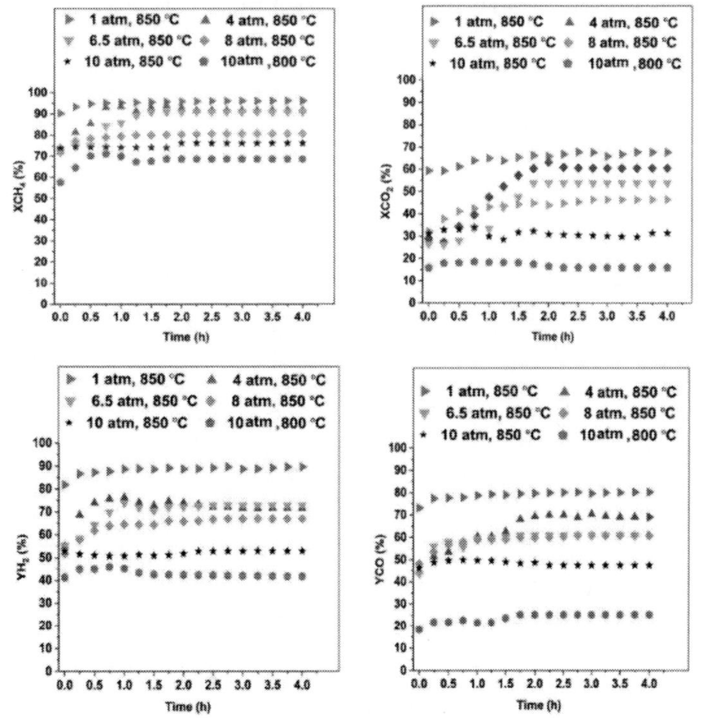

Figure 1. The Ni-UGSO pellet performances in DATR reactions (a) Conversion of CH_4 (%); (b) conversion of CO_2 (%); (c) yield of H_2 (%), (d) yield of CO at 850 °C, feed composition CH_4/CO_2 = 3:1, CH_4/O_2 = 2:1, GHSV: 1000 L/(h*kgcat), P: 1-10 atm, mass of catalyst:300 gm

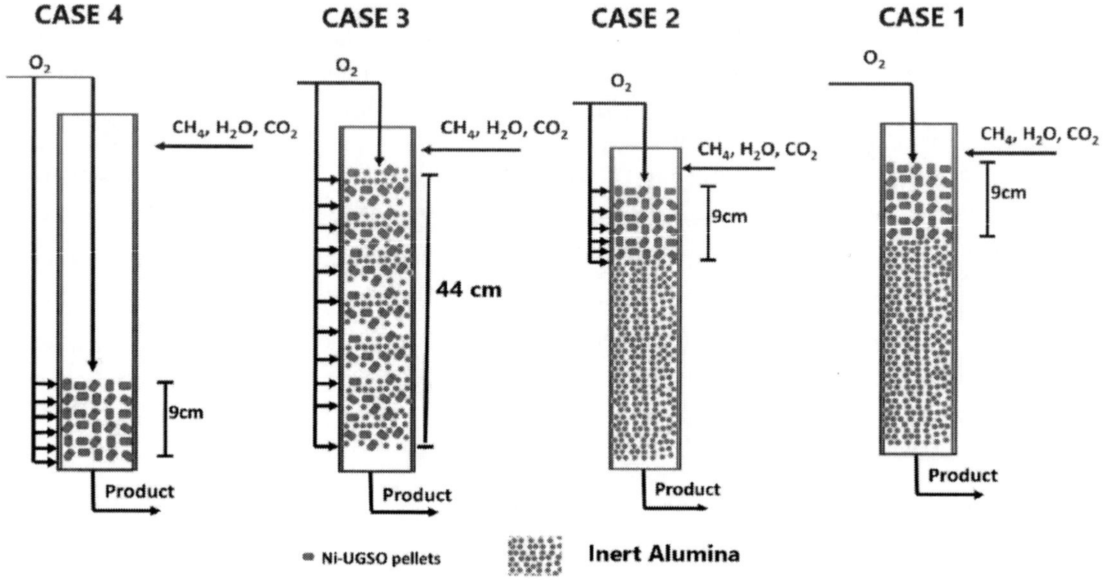

Figure 2. presents a selection scheme for appropriate O_2 circulation and position of the catalytic bed (mass of catalyst:300 gm) in a Kg based Packed bed reactor.

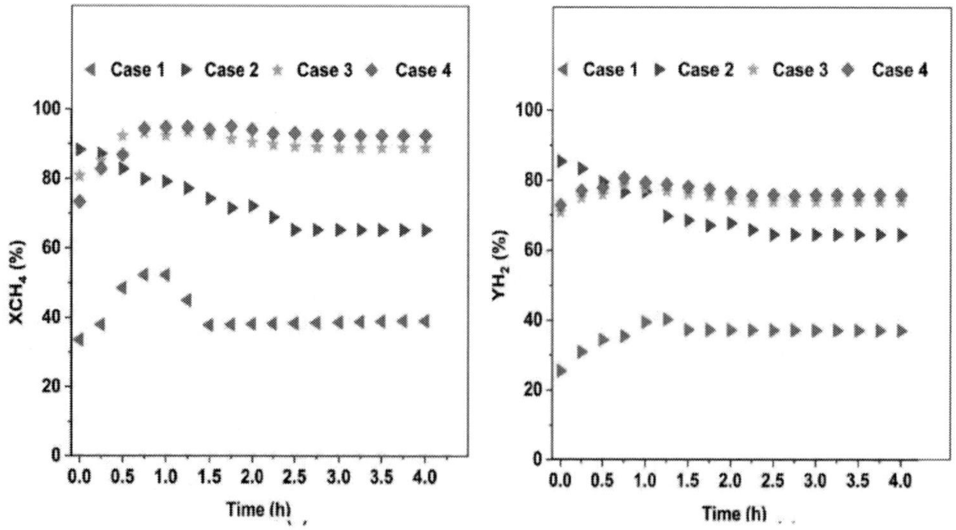

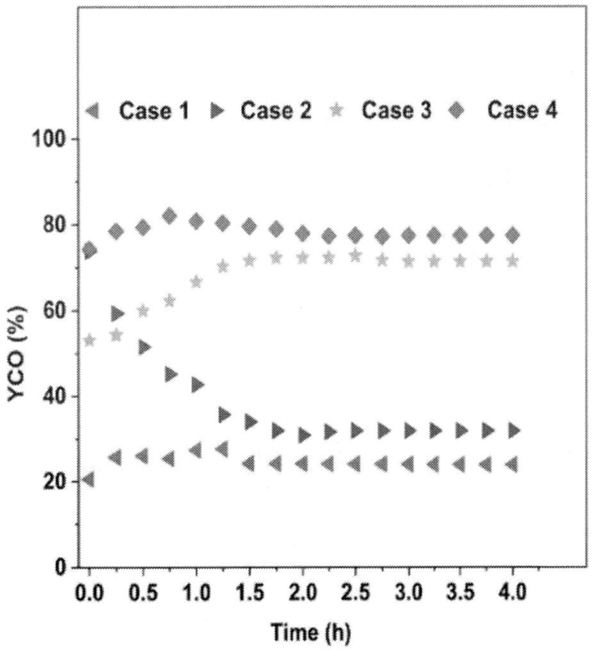

Figure 3. The Ni-UGSO pellet performances in DATR reactions under different cases (a) Conversion of CH_4 (%); (b) yield of H_2 (%), (c) yield of CO at 850 °C, feed composition $CH_4/CO_2 = 3:1$, $CH_4/O_2 = 2:1$, GHSV: 1000 L/(h*kgcat), P: 3.5 atm, mass of catalyst:300 gm.

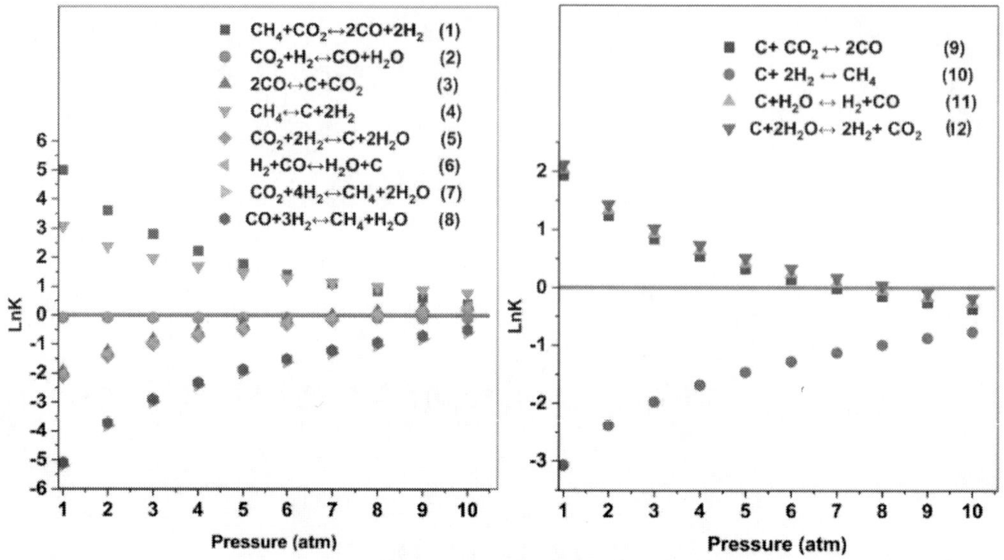

Figure 4. Equilibrium constants, estimated from FactSage as a function of pressure for (a) main DRM and associated side reactions and (b) carbon consumption reactions.

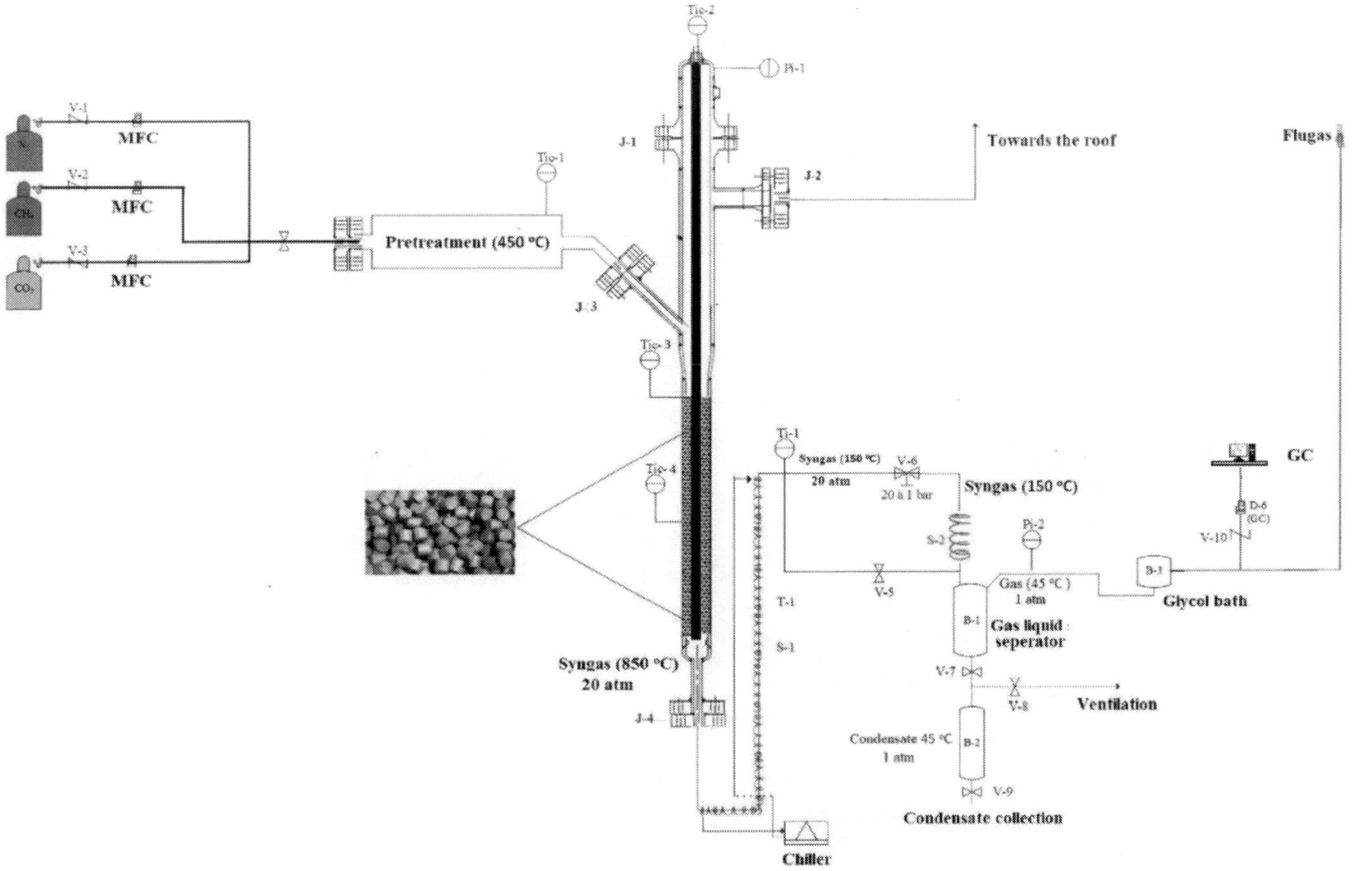

Figure 5. Schematic diagram of the Kg-based packed bed reactor.

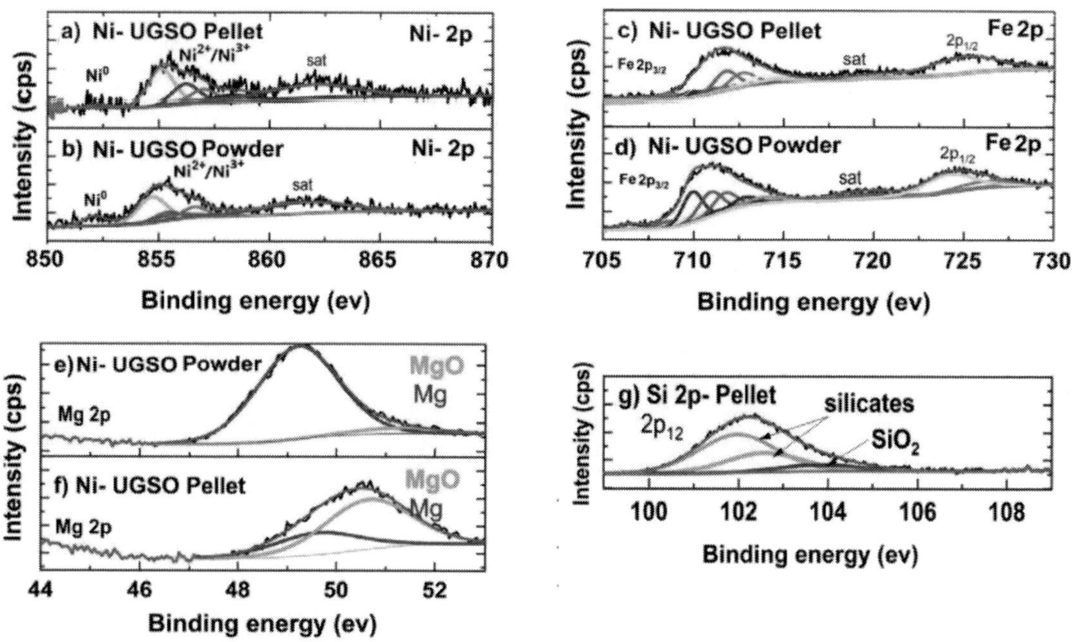

Figure 6. High-resolution XPS spectra of the Ni 2p (a,b), Fe 2p (c,d), Mg 2p (e,f) and Si 2p (g) of Ni-UGSO powder and Pellet sample.

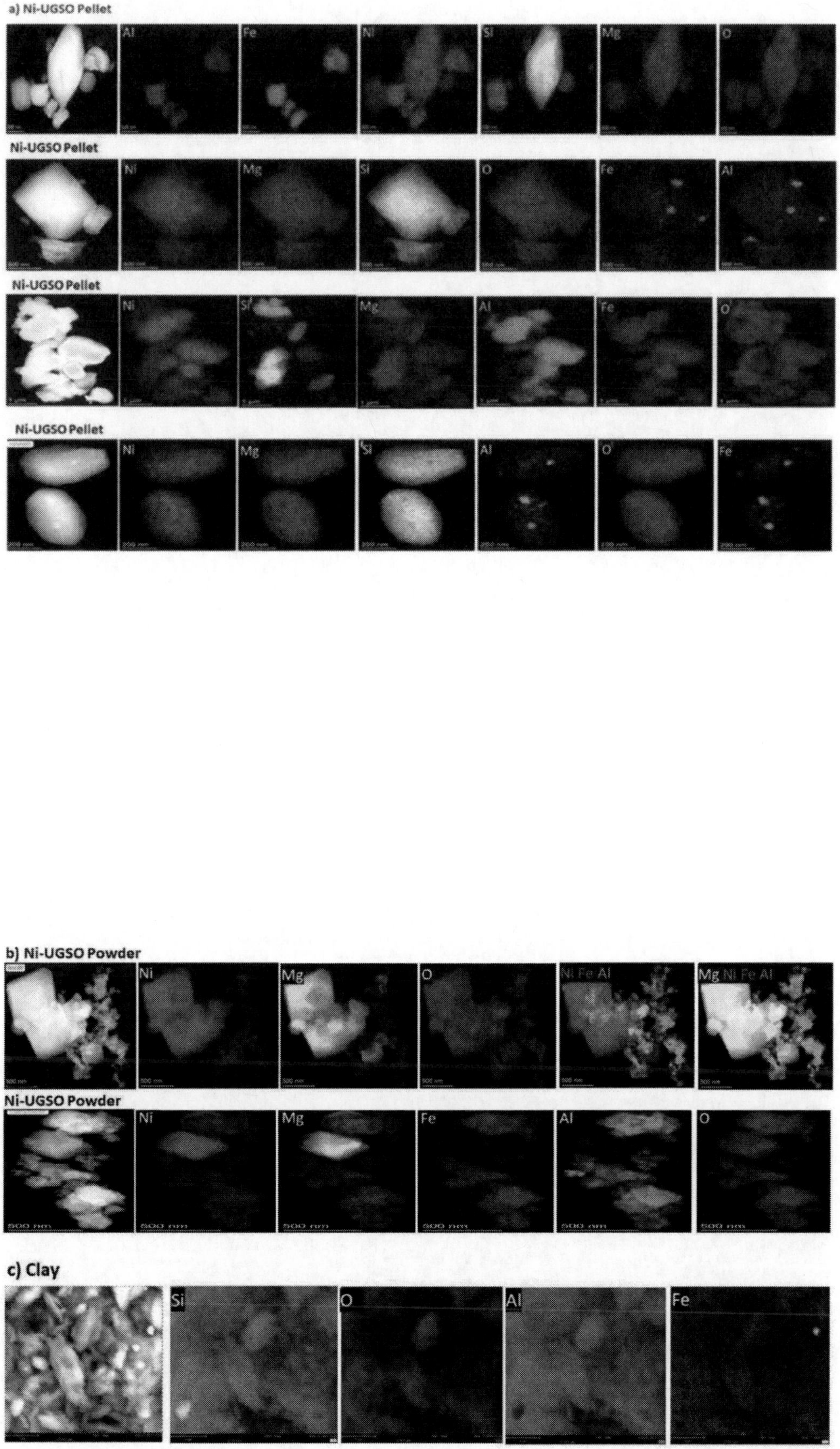

Figure 7. EDX analysis of a) Ni-UGSO pellet, b) Ni-UGSO powder, c) clay

AUTHOR INDEX

AIChE
120 Wall Street, FL 23
New York, NY 10005-4020

CURRAN ASSOCIATES INC.
proceedings.com

ISBN 979-8-3313-1658-7
9 798331 316587